THE COMPLETE AMERICAN MILITARY PILGRIM

THE COMPLETE AMERICAN MILITARY PILGRIM

Guide to 250 battlefields, forts, warships, museums and other places every American military history enthusiast should see

HOWARD KRAMER

Published by **The Complete Pilgrim, LLC**
Marietta, GA

Cover Photo: Lookout Mountain Battlefield,
Chattanooga, Tennessee

Cover Design and Typesetting: JD Smith Design

ISBN 978-1-7325081-4-9 (paperback)
978-1-7325081-5-6 (mobi)

For my grandfathers Edward Kramer and Hank Horowitz;
My uncles Sam, Isadore, Eddie and George, and my aunt Millie
Thank you all for your service to our country

and

In memory of my fellow writer
David Blake

TABLE OF CONTENTS

INTRODUCTION **1**

Methodology

Traveler Cautions

Accuracy of Contents

Concerning the wars against Native Americans

TIMELINE OF AMERICAN MILITARY HISTORY **5**

NEW ENGLAND **15**

1. FORT EDGECOMB STATE HISTORIC SITE, Edgecomb, Maine
2. FORT KNOX HISTORIC SITE, Prospect, Maine
3. FORT AT NUMBER 4 OPEN AIR MUSEUM, Charlestown, New Hampshire
4. BENNINGTON BATTLE MONUMENT, Old Bennington, Vermont
5. BUNKER HILL MONUMENT, Charlestown, Massachusetts
6. USS CONSTITUTION, Charlestown, Massachusetts
7. MASSACHUSETTS 54TH REGIMENT MEMORIAL, Boston, Massachusetts
8. MOUNT AUBURN CEMETERY, Cambridge, Massachusetts
9. MINUTE MAN NATIONAL HISTORICAL PARK, Concord, Massachusetts
10. USS SALEM & UNITED STATES NAVAL SHIPBUILDING MUSEUM, Quincy, Massachusetts
11. USS MASSACHUSETTS & BATTLESHIP COVE MARITIME MUSEUM, Falls River, Massachusetts
12. SPRINGFIELD ARMORY NATIONAL HISTORIC SITE, Springfield, Massachusetts
13. PATRIOT'S PARK, Portsmouth, Rhode Island
14. GREAT SWAMP FIGHT MONUMENT, West Kingston, Rhode Island
15. UNITED STATES COAST GUARD ACADEMY, New London, Connecticut
16. FORT TRUMBULL STATE PARK, New London, Connecticut

NEW YORK 33

17. CASTLE CLINTON NATIONAL MONUMENT & BATTERY PARK MEMORIALS, New York, New York
18. GEORGE M. COHAN MONUMENT, New York, New York
19. USS INTREPID SEA, AIR AND SPACE MUSEUM, New York, New York
20. GENERAL GRANT NATIONAL MEMORIAL, New York, New York
21. WHITE PLAINS BATTLEFIELD SITES, White Plains, New York
22. UNITED STATES MILITARY ACADEMY, West Point, New York
23. BENNINGTON BATTLEFIELD STATE HISTORIC SITE, Walloomsac, New York
24. FORT WILLIAM HENRY MUSEUM, Lake George, New York
25. FORT TICONDEROGA NATIONAL HISTORIC LANDMARK, Ticonderoga, New York
26. SARATOGA NATIONAL HISTORICAL PARK, Stillwater, New York
27. FORT STANWIX NATIONAL MONUMENT, Rome, New York
28. BUFFALO AND ERIE COUNTY NAVAL AND MILITARY PARK, Buffalo, New York

NEW JERSEY & PENNSYLVANIA 47

29. MORRISTOWN NATIONAL HISTORICAL PARK, Morristown, New Jersey
30. MONMOUTH BATTLEFIELD STATE PARK, Manalapan, New Jersey
31. USS NEW JERSEY, Camden, New Jersey
32. TRENTON BATTLE MONUMENT, Trenton, New Jersey
33. WASHINGTON CROSSING PARKS, Titusville, New Jersey & Washington Crossing, Pennsylvania
34. INDEPENDENCE NATIONAL HISTORICAL PARK, Philadelphia, Pennsylvania
35. MUSEUM OF THE AMERICAN REVOLUTION, Philadelphia, Pennsylvania
36. USS OLYMPIA & INDEPENDENCE SEAPORT MUSEUM, Philadelphia, Pennsylvania
37. VALLEY FORGE NATIONAL HISTORICAL PARK, King of Prussia, Pennsylvania
38. AMERICAN MILITARY EDGED WEAPONRY MUSEUM, Intercourse, Pennsylvania
39. NATIONAL CIVIL WAR MUSEUM, Harrisburg, Pennsylvania

40. GETTYSBURG NATIONAL MILITARY PARK, Gettysburg, Pennsylvania
41. FORT NECESSITY NATIONAL BATTLEFIELD, Farmington, Pennsylvania
42. WOODVILLE PLANTATION, Bridgeville, Pennsylvania

MID-ATLANTIC **63**

43. FORT DELAWARE STATE PARK, Delaware City, Delaware
44. FORT MCHENRY NATIONAL MONUMENT, Baltimore, Maryland
45. USS CONSTELLATION, Baltimore, Maryland
46. UNITED STATES NAVAL ACADEMY, Annapolis, Maryland
47. PATUXENT RIVER NAVAL AIR MUSEUM, Lexington Park, Maryland
48. MONOCACY NATIONAL BATTLEFIELD, Frederick, Maryland
49. NATIONAL MUSEUM OF CIVIL WAR MEDICINE, Frederick, Maryland
50. ANTIETAM NATIONAL BATTLEFIELD, Sharpsburg, Maryland
51. DISTRICT OF COLUMBIA WAR MEMORIAL, Washington, DC
52. NAVY-MERCHANT MARINE MEMORIAL, Washington, DC
53. NATIONAL WORLD WAR II MEMORIAL, Washington, DC
54. KOREAN WAR VETERAN'S MEMORIAL, Washington, DC
55. VIETNAM VETERAN'S MEMORIAL & WOMEN'S MEMORIAL, Washington, DC
56. UNITED STATES NAVY MEMORIAL & NAVAL HERITAGE CENTER, Washington, DC
57. NATIONAL MUSEUM OF THE UNITED STATES NAVY, Washington, DC
58. SMITHSONIAN NATIONAL AIR AND SPACE MUSEUM, Washington, DC
59. SMITHSONIAN NATIONAL MUSEUM OF AMERICAN HISTORY, Washington, DC
60. NATIONAL MUSEUM OF AMERICAN JEWISH MILITARY HISTORY, Washington, DC
61. CONGRESSIONAL CEMETERY, Washington, DC
62. ANACOSTIA PARK, Washington, DC
63. THE PENTAGON, Washington, DC

VIRGINIA **85**

64. ARLINGTON NATIONAL CEMETERY, Arlington, Virginia
65. UNITED STATES AIR FORCE MEMORIAL, Arlington, Virginia

66. NATIONAL MUSEUM OF THE MARINE CORPS, Triangle, Virginia
67. MANASSAS NATIONAL BATTLEFIELD PARK, Manassas, Virginia
68. FREDERICKSBURG AND SPOTSYLVANIA NATIONAL MILITARY PARK, Fredericksburg, Virginia
69. CEDAR CREEK AND BELLE GROVE NATIONAL HISTORICAL PARK, Middletown, Virginia
70. JAMESTOWN SETTLEMENT LIVING MUSEUM, Williamsburg, Virginia
71. COLONIAL NATIONAL HISTORICAL PARK, Yorktown, Virginia
72. RICHMOND NATIONAL BATTLEFIELD PARK, Richmond, Virginia
73. PETERSBURG NATIONAL BATTLEFIELD, Prince George, Virginia
74. UNITED STATES ARMY WOMEN'S MUSEUM, Fort Lee, Virginia
75. USS WISCONSIN, Norfolk, Virginia
76. MACARTHUR MEMORIAL MUSEUM, Norfolk, Virginia
77. APPOMATTOX COURT HOUSE NATIONAL HISTORICAL PARK, Appomattox, Virginia
78. NATIONAL CIVIL WAR CHAPLAINS MUSEUM, Lynchburg, Virginia
79. VIRGINIA MILITARY INSTITUTE, Lexington, Virginia
80. HARPER'S FERRY NATIONAL HISTORICAL PARK, Harper's Ferry, West Virginia

MIDWEST **104**

81. PERRYVILLE BATTLEFIELD STATE HISTORIC SITE, Perryville, Kentucky
82. FORT KNOX & GENERAL GEORGE PATTON MUSEUM, Fort Knox, Kentucky
83. NATIONAL MUSEUM OF THE UNITED STATES AIR FORCE, Dayton, Ohio
84. CHARLES YOUNG BUFFALO SOLDIERS NATIONAL MONUMENT, Wilberforce, Ohio
85. PERRY'S VICTORY AND INTERNATIONAL PEACE MEMORIAL, Put-In-Bay, Ohio
86. FALLEN TIMBERS BATTLEFIELD AND FORT MIAMIS NATIONAL HISTORIC SITE, Maumee, Ohio
87. POLAR BEAR EXPEDITION MEMORIAL, Troy, Michigan
88. RIVER RAISIN NATIONAL BATTLEFIELD PARK, Monroe, Michigan
89. YANKEE AIR MUSEUM, Belleville, Michigan
90. INDIANA WORLD WAR MEMORIAL PLAZA, Indianapolis, Indiana
91. USS INDIANAPOLIS NATIONAL MEMORIAL, Indianapolis, Indiana

92. TIPPECANOE BATTLEFIELD PARK, Battle Ground, Indiana
93. NATIONAL VETERAN'S ART MUSEUM, Chicago, Illinois
94. FORT DE CHARTRES STATE HISTORIC SITE, Prairie Du Rocher, Illinois
95. USS COBIA & WISCONSIN MARITIME MUSEUM, Manitowoc, Wisconsin

SOUTH ATLANTIC **120**

96. USS NORTH CAROLINA, Wilmington, North Carolina
97. HANNAH BLOCK HISTORIC USO CENTER, Wilmington, North Carolina
98. MOORES CREEK NATIONAL BATTLEFIELD, Currie, North Carolina
99. FORT RALEIGH NATIONAL HISTORIC SITE, Manteo, North Carolina
100. GUILFORD COURTHOUSE NATIONAL MILITARY PARK, Greensboro, North Carolina
101. FORT SUMTER NATIONAL MONUMENT, Charleston, South Carolina
102. THE CITADEL, Charleston, South Carolina
103. CSS HUNLEY & WARREN LASCH CONSERVATION CENTER, North Charleston, South Carolina
104. USS YORKTOWN & PATRIOT'S POINT, Mount Pleasant, South Carolina
105. KINGS MOUNTAIN NATIONAL MILITARY PARK, Blacksburg, South Carolina
106. COWPENS NATIONAL BATTLEFIELD, Gaffney, South Carolina
107. FORT PULASKI NATIONAL MONUMENT, Savannah, Georgia
108. NATIONAL MUSEUM OF THE MIGHTY EIGHTH AIR FORCE, Pooler, Georgia
109. NATIONAL INFANTRY MUSEUM AND SOLDIER CENTER & FORT BENNING, Columbus, Georgia
110. NATIONAL CIVIL WAR NAVAL MUSEUM, Columbus, Georgia
111. ANDERSONVILLE NATIONAL HISTORIC SITE, Andersonville, Georgia
112. KENNESAW MOUNTAIN NATIONAL BATTLEFIELD PARK, Kennesaw, Georgia
113. CHICKAMAUGA AND CHATTANOOGA NATIONAL MILITARY PARK, Fort Oglethorpe, Georgia

DEEP SOUTH **140**

114. CASTILLO DE SAN MARCOS NATIONAL MONUMENT, St. Augustine, Florida
115. BAY OF PIGS MUSEUM, Miami, Florida
116. OCALA/MARION COUNTY VETERANS MEMORIAL PARK, Ocala, Florida
117. NATIONAL NAVAL AVIATION MUSEUM, Pensacola, Florida
118. SAN JUAN NATIONAL HISTORIC SITE, San Juan, Puerto Rico
119. USS ALABAMA & BATTLESHIP MEMORIAL PARK, Mobile, Alabama
120. TUSKEGEE AIRMEN NATIONAL HISTORIC SITE, Tuskegee, Alabama
121. HORSESHOE BEND NATIONAL MILITARY PARK, Daviston, Alabama
122. AFRICAN AMERICAN MILITARY HISTORY MUSEUM, Hattiesburg, Mississippi
123. BRICE'S CROSSROADS NATIONAL BATTLEFIELD SITE, Guntown, Mississippi
124. TUPELO NATIONAL BATTLEFIELD, Tupelo, Mississippi
125. VICKSBURG NATIONAL MILITARY PARK, Vicksburg, Mississippi
126. MANHATTAN PROJECT NATIONAL HISTORICAL PARK, Oak Ridge, Tennessee
127. SERGEANT ALVIN C. YORK STATE HISTORIC PARK, Pall Mall, Tennessee
128. STONES RIVER NATIONAL BATTLEFIELD, Murfreesboro, Tennessee
129. FRANKLIN BATTLEFIELD, Franklin, Tennessee
130. FORT DONELSON NATIONAL BATTLEFIELD, Dover, Tennessee
131. SHILOH NATIONAL MILITARY PARK, Shiloh, Tennessee
132. NATIONAL WORLD WAR II MUSEUM, New Orleans, Louisiana
133. JEAN LAFITTE NATIONAL HISTORICAL PARK, Chalmette, Louisiana
134. FORT SMITH NATIONAL HISTORIC SITE, Fort Smith, Arkansas
135. PEA RIDGE NATIONAL MILITARY PARK, Garfield, Arkansas

GREAT PLAINS **165**

136. FORT SNELLING, St. Paul, Minnesota
137. FORT ATKINSON STATE PRESERVE, Fort Atkinson, Iowa
138. STARS AND STRIPES MUSEUM AND LIBRARY, Bloomfield, Missouri
139. WILSON'S CREEK NATIONAL BATTLEFIELD, Republic, Missouri
140. NATIONAL WORLD WAR I MUSEUM AND MEMORIAL, Kansas City, Missouri

141. FORT ABERCROMBIE STATE HISTORIC SITE, Abercrombie, North Dakota
142. MINUTEMAN MISSILE NATIONAL HISTORIC SITE, Philip, South Dakota
143. CRAZY HORSE MEMORIAL, Crazy Horse, South Dakota
144. FORT ROBINSON STATE PARK, Crawford, Nebraska
145. FORT LEAVENWORTH, Fort Leavenworth, Kansas
146. BLACK JACK BATTLEFIELD AND NATURE PARK, Wellsville, Kansas
147. FORT SCOTT NATIONAL HISTORIC SITE, Fort Scott, Kansas
148. DWIGHT D. EISENHOWER PRESIDENTIAL LIBRARY AND MUSEUM, Abilene, Kansas
149. FORT RILEY MUSEUMS, Fort Riley, Kansas
150. FORT LARNED NATIONAL HISTORIC SITE, Larned, Kansas
151. FORT SILL & UNITED STATES ARMY ARTILLERY MUSEUM, Fort Sill, Oklahoma
152. WASHITA BATTLEFIELD NATIONAL HISTORIC SITE, Cheyenne, Oklahoma

SOUTHWEST **183**

153. BUFFALO SOLDIERS NATIONAL MUSEUM, Houston, Texas
154. SAN JACINTO BATTLEGROUND STATE HISTORIC SITE, La Porte, Texas
155. BATTLESHIP TEXAS STATE HISTORIC SITE, La Porte, Texas
156. USS LEXINGTON, Corpus Christi, Texas
157. COMMEMORATIVE AIR FORCE AIRPOWER MUSEUM, Midland, Texas
158. PRESIDIO LA BAHIA, Goliad, Texas
159. PALO ALTO BATTLEFIELD NATIONAL HISTORICAL PARK, Brownsville, Texas
160. THE ALAMO MISSION, San Antonio, Texas
161. MILITARY WORKING DOG TEAMS NATIONAL MONUMENT, San Antonio, Texas
162. NATIONAL MUSEUM OF THE PACIFIC WAR, Fredericksburg, Texas
163. FORT DAVIS NATIONAL HISTORIC SITE, Fort Davis, Texas
164. WHITE SANDS MISSILE RANGE MUSEUM, White Sands Missile Range, New Mexico
165. SAND CREEK MASSACRE NATIONAL HISTORIC SITE, Eads, Colorado

166. BENT'S OLD FORT NATIONAL HISTORIC SITE, La Junta, Colorado
167. UNITED STATES AIR FORCE ACADEMY, Air Force Academy, Colorado
168. CHEYENNE MOUNTAIN COMPLEX, Colorado Springs, Colorado
169. COVE FORT HISTORIC SITE, Beaver, Utah
170. FORT BOWIE NATIONAL HISTORIC SITE, Bowie, Arizona
171. DAVIS-MONTHAN AIR FORCE BASE AIRCRAFT BONEYARD, Tucson, Arizona
172. FORT CHURCHILL STATE HISTORIC PARK, Silver Springs, Nevada

NORTHWEST AND ALASKA **205**

173. FORT LARAMIE NATIONAL HISTORIC SITE, Fort Laramie, Wyoming
174. LITTLE BIGHORN BATTLEFIELD NATIONAL MONUMENT, Crow Agency, Montana
175. BIG HOLE NATIONAL BATTLEFIELD, Wisdom, Montana
176. FORT HALL, Pocatello, Idaho
177. FORT CLATSOP, Astoria, Oregon
178. FORT YAMHILL BLOCKHOUSE, Dayton, Oregon
179. FORT VANCOUVER NATIONAL HISTORIC SITE, Vancouver, Washington
180. ALEUTIAN WORLD WAR II NATIONAL HISTORIC AREA, Unalaska, Alaska

CALIFORNIA & PACIFIC ISLANDS **214**

181. ALCATRAZ ISLAND, San Francisco, California
182. FORT POINT NATIONAL HISTORIC SITE, San Francisco, California
183. USS PAMPANITO & SAN FRANCISCO MARITIME NATIONAL HISTORICAL PARK, San Francisco, California
184. WEST COAST MEMORIAL TO THE MISSING, San Francisco, California
185. ABRAHAM LINCOLN BRIGADE MONUMENT, San Francisco, California
186. USS HORNET MUSEUM SHIP, Alameda, California
187. ROSIE THE RIVETER HOME FRONT NATIONAL HISTORICAL PARK, Richmond, California
188. PORT CHICAGO NAVAL MAGAZINE NATIONAL MEMORIAL, Concord, California
189. TULE LAKE WAR RELOCATION CENTER, Tule Lake, California
190. UNITED STATES NAVY SEABEE MUSEUM, Port Hueneme, California
191. USS IOWA MUSEUM SHIP, Los Angeles, California

192. NORTHWOOD GRATITUDE AND HONOR MEMORIAL, Irvine, California
193. SAN DIEGO AIR AND SPACE MUSEUM, San Diego, California
194. USS MIDWAY MUSEUM, San Diego, California
195. BOB HOPE MEMORIAL, San Diego, California
196. USS MISSOURI MEMORIAL, Honolulu, Hawaii
197. USS ARIZONA MEMORIAL, Honolulu, Hawaii
198. USS BOWFIN SUBMARINE MUSEUM, Honolulu, Hawaii
199. WAR IN THE PACIFIC NATIONAL HISTORICAL PARK, Piti, Guam
200. AMERICAN MEMORIAL PARK, Garapan, Northern Mariana Islands

FOREIGN - NORTH AMERICA AND CARIBBEAN 234
201. MONUMENT TO THE VICTIMS OF THE USS MAINE, Havana, Cuba
202. SAN JUAN HILL BATTLEFIELD & SANTIAGO SURRENDER TREE, Santiago de Cuba, Cuba
203. CHAPULTEPEC PARK, Mexico City, Mexico
204. MEXICO CITY NATIONAL CEMETERY, Mexico City, Mexico
205. FORT MALDEN NATIONAL HISTORIC SITE, Amherstburg, Canada
206. THAMES BATTLEFIELD, Chatham-Kent, Canada

FOREIGN - NORTHERN EUROPE 241
207. AMERICAN AIR MUSEUM IN BRITAIN, Cambridge, United Kingdom
208. LUSITANIA MEMORIAL GARDEN, Kinsale, Ireland
209. UNITED STATES NAVAL MONUMENT AT BREST, Brest, France
210. NORMANDY AMERICAN CEMETERY AND MEMORIAL, Colleville-sur-Mer, France
211. UTAH BEACH AMERICAN MEMORIAL & MUSEUM, Ste-Marie-du-Mont, France
212. MEMORIAL MUSEUM OF THE BATTLE OF NORMANDY, Bayeux, France
213. TOURS AMERICAN MONUMENT, Tours, France
214. CHAUMONT AMERICAN EXPEDITIONARY FORCE HEADQUARTERS MARKER, Chaumont, France
215. LAFAYETTE ESCADRILLE MEMORIAL CEMETERY, Marnes-la-Coquette, France
216. BELLEAU WOOD AMERICAN MONUMENT, Belleau, France

217. CHATEAU-THIERRY AMERICAN MONUMENT, Chateau-Thierry, France
218. BELLICOURT AMERICAN MONUMENT, Bellicourt, France
219. MEUSE-ARGONNE AMERICAN MEMORIAL & MONTFAUCON AMERICAN MONUMENT, Montfaucon-d'Argonne, France
220. LUXEMBOURG AMERICAN CEMETERY AND MEMORIAL, Luxembourg City, Luxembourg
221. BASTOGNE WAR MUSEUM & MARDASSON MEMORIAL, Bastogne, Belgium
222. ARDENNES AMERICAN CEMETERY AND MEMORIAL, Neupre, Belgium
223. NETHERLANDS AMERICAN CEMETERY AND MEMORIAL, Margraten, Netherlands
224. HURTGEN FOREST BATTLEFIELD, Hurtgenwald, Germany
225. PEACE MUSEUM BRIDGE AT REMAGEN, Remagen, Germany
226. ALLIED MUSEUM, Berlin, Germany
227. CHECKPOINT CHARLIE MUSEUM, Berlin, Germany
228. MUSEUM OF THE PRISONER OF WAR CAMPS, Zagan, Poland
229. BIG THREE MONUMENT, Yalta, Russia

FOREIGN – SOUTHERN EUROPE & NORTH AFRICA **264**

230. ANZIO BEACHHEAD MUSEUM, Anzio, Italy
231. SICILY-ROME AMERICAN CEMETERY AND MEMORIAL, Neptuno, Italy
232. FLORENCE AMERICAN CEMETERY AND MEMORIAL, Tavarnuzze, Italy
233. NAVAL MONUMENT AT GIBRALTAR, Gibraltar
234. WESTERN NAVAL TASK FORCE MARKER, Casablanca, Morocco
235. NORTH AFRICA AMERICAN CEMETERY AND MEMORIAL, Carthage, Tunisia

FOREIGN – FAR EAST **270**

236. UNITED NATIONS MEMORIAL CEMETERY & KOREAN WAR MONUMENT, Busan, South Korea
237. LEGATION QUARTER, Beijing, China
238. JIANGSHAN RAIDER MEMORIAL HALL, Bao'an, China
239. HUE WAR MUSEUM, Thura Thien Hue, Vietnam
240. KHE SANH COMBAT BASE, Huong Hoa, Vietnam

241. HIROSHIMA PEACE MEMORIAL PARK & MUSEUM, Hiroshima City, Japan

242. CORNERSTONE OF PEACE, Itoman, Japan

243. MOUNT SURIBACHI MEMORIAL, Iwo Jima, Japan

244. MOUNT SAMAT NATIONAL SHRINE, Pilar, Philippines

245. CLARK VETERANS CEMETERY, Mabalacat, Philippines

246. CABANATUAN AMERICAN MEMORIAL, Cabanatuan City, Philippines

247. MANILA AMERICAN CEMETERY AND MEMORIAL, Taguig City, Philippines

248. MACARTHUR LANDING MEMORIAL NATIONAL PARK, Palo, Philippines

249. MARKER AT PAPUA NEW GUINEA, Port Moresby, Papua New Guinea

250. GUADALCANAL AMERICAN MEMORIAL, Honiara, Solomon Islands

COMING SOON **286**

NATIONAL MUSEUM OF THE UNITED STATES ARMY, Fort Belvoir, Virginia

NATIONAL MEDAL OF HONOR MUSEUM, Mount Pleasant, South Carolina

INDEX OF SITES BY TYPE **288**

INDEX OF SITES BY CONFLICT **305**

About The Author **317**

INTRODUCTION

For those of you familiar with my previous books, you are no doubt aware of my interest in spiritual travel. I have been making pilgrimages to the great religious sites of the world for the better part of half of my life. However, places of worship are not my only interest, especially when it comes to historic sightseeing. I have a true passion for history, and I love to visit places where the great moments of history have taken place.

For better or for worse, many of the most important moments in history have been related to war. Few things have impacted the direction of humanity as much as warfare, and the world is scarred with countless battlefields to prove it. Monuments around the globe commemorate those who fought and died, while museums preserve the stories and artifacts of mankind's most enduring and terrible pastime.

Over the course of nearly three decades I have visited many places of military interest, both in the United States and abroad. With the recent publication of *The Complete American Pilgrim* and *The Complete Christian Pilgrim*, I am taking a break from religion to write about military tourism. *The Complete American Military Pilgrim* is the culmination of three years of writing and many years of travel.

The Complete Pilgrim was introduced to the world in 2014 as an online travel blog (www.thecompletepilgrim.com). It now receives over a hundred thousand visitors annually. Not huge, but decent by travel blog standards. It is my hope that between the website and books I am engendering my readers with a greater interest in a love for history, travel and a desire to more fully understand the world in which we live.

Methodology

The process of selecting 250 must-see sites of American military interest was not nearly as involved as picking out the places for my religious travel books. Nevertheless in my research I identified more than five hundred worthwhile places to choose from, both inside and outside the United States. From battlefields to forts to ships to monuments to museums, and everything else in between, there were plenty of options. In order to narrow down the field, four factors were considered for each location:

1) Is the site already popular among military history enthusiasts, with a substantial number of annual visitors; 2) Did the site play a critical role in American military history; 3) Is anybody buried at the site of particular military importance; 4) Is there a collection of exceptional artifacts on display at the location.

Obviously this was a highly subjective process. I truly hope that I did not underemphasize any particularly worthy location, or worse, overlook one altogether. As with my other books, it was difficult to organize these entries based on time period or theme. So once again the entries are organized by geography. However, there are indexes in the back that list sites by category and by conflict.

Traveler Cautions

The majority of the sites in this book are run by either the National Park Service or the American Battle Monuments Commission. Because of this there is generally little problem in terms of visitor accessibility. Most of the sites herein are easily reached by motor transportation, or even on foot in some major cities, and the majority of places that are open to the public are handicapped accessible. Very few of the places listed are in dangerous areas, but travelers should always take basic safety precautions.

When this book was in production, every effort was made to insure that all visitor information was as accurate and up-to-date as possible, and all information contained herein was deemed reliable as of the completion of writing in 2020. However, *The Complete American Military Pilgrim* and its writers are not responsible for any changes in schedules, opening times or contact information. Please note also that some sites, particularly museums, do occasionally move to new locations. Readers of this book are advised to verify any and all visitor information prior to travel.

Accuracy of Contents

Every effort was made to ensure the accuracy of information about each and every site in this book. As of the time of this writing, the author had personally visited approximately sixty of these destinations and wrote based on first-hand observations supplemented as needed by research. For those places not personally visited, writing was based on thorough research from reliable sources both online and in print.

If any part of this book is found to be in error, or some information of significance has been omitted, *The Complete Pilgrim* encourages its readers to reach out to us at our website (www.thecompletepilgrim.com) with any suggestions. Upon confirmation, changes may be made accordingly at the discretion of the author. A new edition of this volume will likely be issued sometime in 2023.

Concerning the wars against Native Americans

Finally, a note about the European and American settler wars against Native American tribes. This is obviously a very sensitive subject. While the battlefields of the American Revolution, the American Civil War and the World Wars can inspire a sense of pride in visitors, not so the battlefields of the wars against Native American tribes. These wars represent a dark stain on American history, as did most European conquests throughout the world during the colonial era. Nevertheless they are a part of America's history, and by their nature merited inclusion.

The wars against Native Americans raged for nearly three centuries, from the earliest clashes in the 1630s to the final last stands of the early 1900s. The United States and its colonial predecessors fought nearly a hundred named wars against hundreds of tribes from the Atlantic to the Pacific. Some of these conflicts were brief and consisted only of minor skirmishes, while others lasted for decades and witnessed major battles. Many of these wars involved the wholesale slaughter of native tribes, and most of them resulted in the loss of tribal territory.

More than two dozen locations that played some part in the Native American wars are included in this book. Some of these were the locations of true battles, while others were in fact actually massacres.

In general these places are now treated as memorials to the local tribes who fought and died in defense of their lands. These battles are now regarded by many people as part of a systematic act of genocide. Visitors to such battlefields should understand that these places are viewed by *The Complete Pilgrim* as places of commemoration for the victims.

TIMELINE OF AMERICAN MILITARY HISTORY

COLONIAL ERA (1688-1783)

1675 – ***King Philip's War*** (1675-1678)

1688 - ***King William's War*** (1688-1697)

1702 - ***Queen Anne's War*** (1702-1713)

1744 - ***King George's War*** (1744-1763)

1754 - ***French and Indian War*** (1754-1763)

- 1754 - Battle of Fort Necessity
- 1755 – Siege of Fort William Henry, Battle of Lake George
- 1759 – Battle of the Plains of Abraham

1775 - ***American Revolution*** (1775-1783)

- 1775 – Continental Army and Navy established by act of Continental Congress
- 1775 – Battle of Lexington and Concord; Battle of Bunker Hill
- 1776 – Declaration of Independence signed at Philadelphia
- 1776 – Battle of Moores Creek Bridge, Battle of Long Island, Battle of White Plains, Battle of Trenton
- 1777 – Battle of Bennington, Battle of Saratoga, Winter at Valley Forge (ends 1778)
- 1778 – Battle of Rhode Island, Battle of Monmouth
- 1780 – Battle of Kings Mountain

1781 – Battle of Cowpens, Battle of Guilford Courthouse, Siege of Yorktown; Surrender of Cornwallis

1783 – American independence recognized by Treaty of Paris

MANIFEST DESTINY ERA & EARLY AMERICAN INDIAN WARS (1784-1860)

1776 - ***Cherokee War*** (1776-1795)

1784 – United States Army established by act of Congress

1785 - ***Northwest Indian War*** (1785-1794)

1794 – Battle of Fallen Timbers

1786 - ***Shay's Rebellion*** (1786-1787)

1791 - ***Whiskey Rebellion*** (1791-1794)

1791 – Skirmish at Woodville Plantation

1794 – Permanent Naval force established by act of Congress

1798 – United States Marine Corps established by act of Congress

1798 - ***Quasi War with France*** (1798-1800)

1801 - ***First Barbary War*** (1801-1805)

1803 – Battle of Tripoli Harbor

1802 – United States Military Academy established at West Point, New York

1811 - ***Tecumseh's War*** (1811-1813)

1811 – Battle of Tippecanoe

1812 - ***War of 1812*** (1812-1815)

1813 – Battle of Lake Erie, Battle of Frenchtown, Battle of the Thames

1814 – Bombardment of Fort McHenry; "Star Spangled Banner" written by Francis Scot Key

1815 – Battle of New Orleans

1813 - ***Creek War*** (1813-1814)

1814 – Battle of Horseshoe Bend

1815 - ***Second Barbary War*** (1815)

1817 - ***First Seminole War*** (1817-1818)

1832 – ***Sumatran Expeditions*** (1832-1839)

1832 – ***Black Hawk War*** (1832)

1835 - ***Texas Revolution*** (1835-1836)

1836 – Siege of the Alamo, Battle of San Jacinto

1835 - ***Second Seminole War*** (1835-1842)

1838 - ***Aroostook War*** (1838-1839)

1845 – United States Naval Academy established at Annapolis, Maryland

1846 - ***Mexican-American War*** (1846-1848)

1846 – Battle of Palo Alto, Battle of Resaca de la Palma

1847 – Battle of Chapultepec

1854 - ***Bleeding Kansas*** (1854-1861)

1856 – Battle of Black Jack

1855 - ***Third Seminole War*** (1855-1858)

1855 – ***Fiji Expeditions*** (1855-1859)

1855 - ***Yakima, Puget Sound and Rogue River Wars*** (1855-1858)

1856 - ***Second Opium War*** (1856-1859)

1857 - ***Mormon Rebellion*** (1857-1858)

1860 - ***Navajo War*** (1860-1864)

AMERICAN CIVIL WAR (1861-1865)

1859 – Raid on Harper's Ferry (pre-conflict)

1861 - ***American Civil War*** (1861-1865)

1861 – Siege of Fort Sumter, First Battle of Bull Run, Battle of Wilson's Creek; "Battle Hymn of the Republic" written by Julia Ward Howe; "Stars and Stripes" newspaper founded

1862 – East - Second Battle of Bull Run, Battle of Antietam, Battle of Fredericksburg

1862 – West - Battle of Fort Donelson, Battle of Pea Ridge, Battle of Shiloh, Battle of Stones River

1863 – East - Battle of Chancellorsville, Battle of Gettysburg

1863 – West - Siege of Vicksburg, Battle of Chattanooga, Battle of Chickamauga

1864 – East – Battle of the Wilderness, Battle of Spotsylvania Courthouse, Battle of Cold Harbor, Battle of Monocacy, Battle of the Crater, Battle of Cedar Creek, Siege of Petersburg (ends 1865)

1864 – West – Battle of Kennesaw Mountain, Battle of Tupelo, Battle of Franklin

1864 – Arlington National Cemetery established

1865 – Battle of Fort Blakeley, Battle of Appomattox Courthouse; Surrender of Robert E. Lee

1865 – Battle of Palmito Ranch (post-conflict)

IMPERIAL ERA & LATE AMERICAN INDIAN WARS (1866-1913)

1849 - ***Apache Wars*** (1849-1886)

1854 - ***Sioux Wars*** (1854-1891)

- 1868 – Washita Massacre
- 1876 – Great Sioux War (ends 1877); Battle of Little Bighorn
- 1877 – Battle of the Big Hole
- 1890 – Wounded Knee Massacre

1860 – ***Paiute War***

1863 - ***Colorado War*** (1863-1865)

- 1864 – Sand Creek Massacre

1864 - ***Snake War*** (1864-1868)

1866 – Ulysses S. Grant named the first General of the Army

1867 – ***Comanche Campaign*** (1867-1875)

1869 – William T. Sherman becomes General of the Army

1871 – ***Korean Expedition***

1872 – ***Modoc War*** (1872-1873)

1874 - ***Red River War*** (1874-1875)

1877 – ***Nez Perce War***

1878 – ***Bannock War***

1878 – ***Cheyenne War*** (1878-1879)

1879 – ***Sheepeater Indian War***

1888 – Philip Sheridan becomes General of the Army

1897 – "Stars and Stripes Forever" written by John Philip Sousa

1898 - ***Spanish-American War*** (1898)

- 1898 – Battle of San Juan Hill, Battle of Manila Bay

1898 - ***Second Somoan Civil War*** (1898-1899)

1899 - ***Boxer Rebellion*** (1899-1901)

1900 – Siege of the International Legations

1899 - ***Philippine-American War*** (1899-1902)

1907 – "Anchors Away" adopted as the anthem of the United States Navy

WORLD WAR I & INTERWAR PERIOD (1914-1938)

1914 - ***World War I*** (1914-1918)

1915 – American flyers fight for France in the Lafayette Escadrille; Sinking of the Lusitania

1917 – "Over There" written by George M. Cohan

1918 – Battle of Belleau Wood, Second Battle of the Marne, Battle of St. Quentin Canal, Meuse-Argonne Offensive

1918 - ***Russian Civil War*** (1918-1920)

1919 – Polar Bear Expedition

1923 – ***Posey War***

1929 – "Marine's Hymn" adopted as the anthem of the United States Marine Corps

1930 – United States Coast Guard Academy established at New London, Connecticut

1932 – Veteran's Bonus March on Washington DC

1936 – ***Spanish Civil War*** (1936-1939)

1937 – American volunteers fight in the Lincoln Brigade

WORLD WAR II (1939-1945)

1939 - ***World War II*** (1939-1945)

1941 – Pacific – Battle of Pearl Harbor, Battle of Guam, First Philippines campaign (ends 1942), Battle of Bataan

1941 – Other - United Service Organizations (USO) founded; Tuskegee Airmen formed

1942 – Europe – Invasion of French North Africa

1942 – Pacific – Doolittle Raid, Battle of the Coral Sea, Battle of Midway, Aleutian Islands campaign (ends 1943), Guadalcanal campaign (ends 1943), New Guinea campaign (ends 1945)

1942 – Other - Japanese Americans interred in concentration camps (ends 1946); Manhattan Project initiated

1943 – Europe – Battle of Kasserine Pass, Invasion of Sicily, Invasion of Italy

1943 – Pacific – Solomon Islands campaign

1943 – Other - Headquarters for the Department of Defense opened at the Pentagon; Tehran Conference

1944 – Europe – Battle of Monte Cassino, Battle of Anzio, Invasion of Normandy, Advance to the Rhine, Battle of Aachen, Battle of Hurtgen Forest, Battle of the Bulge (ends 1945), Siege of Bastogne

1944 – Pacific – Second Philippines campaign (ends 1945), Battle of Leyte

1944 – Other - George Marshall, Douglas MacArthur, Dwight Eisenhower and Henry Arnold named Generals of the Army; William Leahy, Ernest King and Chester Nimitz Named Fleet Admirals; Port Chicago disaster

1945 – Europe - Battle of Remagen

1945 – Pacific – Battle of Iwo Jima, Battle of Okinawa, Bombing of Hiroshima and Nagasaki

1945 – Other - Atomic Bomb tested at White Sands; Yalta Conference

POSTWAR ERA (1946-PRESENT)

1947 – United States Air Force established by act of Congress; "Off We Go Into The Wild Blue Yonder" adopted as the anthem of the United States Air Force

1950 - ***Korean War*** (1950-1953)

1950 – Battle of the Pusan Perimeter, Battle of Incheon

1954 – United States Air Force Academy established at Colorado Springs, Colorado

1956 – "The Army Goes Rolling Along" adopted as the anthem of the United States Army

1958 - ***Lebanon Crisis*** (1958)

1961 - ***Bay of Pigs Invasion*** (1961)

1965 - ***Dominican Civil War*** (1965-1966)

1965 - ***Vietnam War*** (1965-1975)

1968 – Battle of Hue, Battle of Khe Sanh

1966 – Cheyenne Mountain complex opened in Colorado

1967 - ***Cambodian Civil War*** (1967-1975)

1982 - ***Intervention in Lebanon*** (1982-1984)

1983 - ***Invasion of Grenada*** (1983)

1987 – "The Stars and Stripes Forever" adopted as the national march of the United States

1989 - ***Invasion of Panama*** (1989-1990)

1990 - ***Gulf War*** (1990-1991)

1992 - ***Intervention in Somalia*** (1992-1995)

1992 - ***Bosnia War*** (1992-1995)

2001 – Terrorist Attacks of 9/11

2001 - ***Afghanistan War*** (2001-present)

2003 – ***Iraq War*** (2003-2011)

2014 - ***War against the Islamic State*** (2014-present)

NEW ENGLAND

1. FORT EDGECOMB STATE HISTORIC SITE

66 Fort Road, Edgecomb, Maine, 04556

Site Type: Wooden Blockhouse
Conflict: War of 1812
Dates: Originally completed in 1809
Designations: National Register of Historic Places, Fort Edgecomb National Historic District, Maine State Historic Site
Web: www.maine.gov/cgi-bin/online/doc/parksearch/details.pl?park_id=32 (official website)

Fort Edgecomb is one of the oldest surviving coastal fortifications in the United States. Constructed during the turbulent era of the Napoleanic Wars then raging in Europe, Fort Edgecomb was used to defend this important shipbuilding area from both the English and the French, and to enforce economic embargoes against both. The fort was most heavily in service during the War of 1812 and served as a deterrent to British raids in New England.

During the early 19th century the ongoing conflict between France and Britain began to impact neutral shipping in the Atlantic, threatening to drag America into the war, which eventually it did in 1812. To counter this threat America built a series of coastal fortifications to defend key ports and ship building areas, especially in New England. Fort Edgecomb, completed in 1809, was one of these.

Fort Edgecomb was designed not only to defend the Maine coast but also to enforce Thomas Jefferson's Embargo Act passed the year before. The fort was brought into active service during the War of 1812, and though it never saw action, it was used as a detention facility for British prisoners of war. Abandoned in 1818, Edgecomb was briefly brought into service once more during the American Civil War. Fort Edgecomb was listed on the National Register of Historic Places in 1969 and named a National Historic District in 1991.

Fort Edgecomb is one of the best preserved early American coastal forts and is home to a rare surviving octagonal 19th century blockhouse. The main point of interest is the blockhouse, at the highest point of the fort, which was built to use both cannons and muskets in the fort's defense. Further down the hill are the surviving earthworks where the main battery guarding the shoreline was located.

---

2. FORT KNOX HISTORIC SITE

740 Fort Knox Road, Prospect, Maine, 04981

Site Type: Masonry Fort
Dates: Mostly completed in 1869 (never finished)
Designations: National Register of Historic Places, National Historic Landmark, Maine State Historic Site
Web: www.maine.gov/mdot/pnbo/fortknox (official website)

Fort Knox is one of the largest surviving 19th coastal fortifications in New England. Mostly intact, it was built to strengthen Maine's maritime defenses in the aftermath of the War of 1812. Unfortunately Fort Knox was already obsolete before it was even completed, and construction was abandoned before it was finished. The fort was manned during the American Civil War but never saw service in combat.

During the American Revolution and War of 1812, Maine had proven extremely vulnerable to attacks by the British navy. In the 1830s, when the United States and Great Britain were at odds over the border between Maine and Canada, it was decided that new fortifications were required to protect American interests in New England. Construction on Fort Knox began in 1844.

Although tensions with Britain lessened over time, work on the fort continued for over two decades. It was manned during both the American Civil War and the Spanish-American War, but neither conflict even came close to threatening the Maine coast. The fort was eventually decommissioned and sold to the state of Maine in 1923. Fort Knox was listed on the National Register of Historic Places in 1969 and named a National Historic Landmark in 1970.

Fort Knox is a well-preserved 19th century fortification despite the fact that it was never completed. This is due to the fact that first, the bulk of the fort is constructed from granite; and second, it was never damaged in combat. Much of the fort looks as it did in the mid-19th century, although the artillery pieces on display cover a range of military eras. Of particular interest is the shot furnace, one of the last surviving structures of its kind in the United States.

---

3. FORT AT NUMBER 4 OPEN AIR MUSEUM

267 Springfield Road, Charlestown, New Hampshire, 03603

Site Type: Wooden Stockade Fort
Conflicts: King George's War, French and Indian War, American Revolution
Dates: Originally completed in 1743
Web: www.fortat4.org (official website)

The Fort at Number 4 was a small fortification that nominally marked the northern end of British settlement in New Hampshire for a period during the mid-18th century. It was fought over during both King George's War and the French and Indian War, and was a gathering point for patriot militia during the American Revolution. The site is currently maintained as a living history museum.

Settlers from the British colonies began arriving in the area around what is now Charlestown in the 1730s. Due to the proximity of French Canada and the possibility of war between France and England, the settlers constructed a small fort here to defend against raids. During King George's War the fort endured several attacks, and though abandoned for a time remained under British control.

During the French and Indian War a few years later the Fort at Number 4 was once again active. It became an important British strongpoint and suffered from French raids during the conflict. At one point Robert Rogers of Rogers' Rangers fame made the fort a base of operations. During the early years of the American Revolution the remains of the fort became a marshalling location for the New Hampshire militia. The fort was partially restored and repurposed as a living museum in the 20th century.

The Fort at Number 4 Museum is an excellent example of a Colonial era wooden stockade fortification, despite the fact that it consists primarily of reconstruction work. The stockade, with gaps large enough to shoot through, encloses the perimeter. Inside a thick cluster of log houses forms a second defensive ring around the small mustering yard. The museum on sight has exhibits which include artifacts from the original fort era. Costumed hosts offer tours and talks about life in the fort, and occasionally there are live military displays.

---

4. BENNINGTON BATTLE MONUMENT

15 Monument Circle, Old Bennington, Vermont, 05201

Site Type: Monument – Battle of Bennington
Conflict: American Revolution
Dates: Battle fought on August 16, 1777; Monument dedicated in 1891
Designations: National Register of Historic Places, Bennington National Historic District, Vermont State Historic Site
Web: www.benningtonbattlemonument.com (official website)

The Battle of Bennington was a decisive victory for the American forces during the early stages of the Revolution, and one that contributed directly to the eventual British collapse in the Northern colonies. While the Bennington Battle Monument is located in Vermont, the site of the battle was actually about ten miles away in New York. The field where the engagement took place is now the Bennington Battlefield State Historic Site.

The town of Bennington was the objective of a Hessian force seeking the arms and supplies that were being stored in the town. However, the Hessians were stopped by an army of New Englanders before they got anywhere near Bennington. The monument, dedicated in 1891, stands near the site where the military stores were kept. The Bennington Battle Monument was listed on the National Register of Historic Places in 1971and included in the Bennington National Historic District in 1984.

The Bennington Battle Monument is a massive granite obelisk inside of which is a museum with exhibits on the campaign. An elevator gives public access to the observatory at the top. Outside the base of the obelisk are statues commemorating the American commanders, including John Stark. Stark, the hero of the battle, became famous for his declaration of "Live Free or Die", which is now the motto of state of New Hampshire.

---

5. BUNKER HILL MONUMENT

Monument Square, Charlestown, Massachusetts, 02129

Site Type: Battlefield – Battle of Bunker Hill
Conflict: American Revolution
Dates: Battle fought on June 17, 1775; Monument dedicated in 1843
Designations: National Register of Historic Places, National Historic Landmark, National Historical Park
Web: www.nps.gov/bost/planyourvisit/bhm.htm (official website)

The Bunker Hill Monument commemorates the site of the Battle of Bunker Hill, one of the most storied military engagements in American history. It is part of the Boston National Historical Park, a collection of several sites in the city of Boston that are largely connected to the Revolutionary War. These include the Siege of Boston and the Battle of Lexington and Concord. Many of these places are located along the city's famous Freedom Trail.

In the years leading up to the American Revolution Boston was the epicenter of anti-British sentiment, and activity, in the colonies. Political meetings took place at Faneuil Hall, the Old South Meeting House where the Boston Tea Party was organized, and the Old State House where the Boston Massacre took place. On April 18, 1775 lanterns hanging from the bell tower of the Old North Church signaled the beginning of the Battle of Lexington and Concord.

A few months later a large British force from Boston marched to Breed's Hill where American rebels held a strongly fortified position. The ensuing misnamed Battle of Bunker Hill was a victory for the British, but a pyrrhic one. The British successfully drove the Americans from the hill but their casualties were so high they were unable to follow up the victory in any meaningful way. The Bunker Hill Monument and several other locations throughout the city were collectively designated the Boston National Historical Park in 1974.

The Bunker Hill Monument, which is also independently a National Historic Landmark, is a massive granite obelisk that crowns the hill on which the Battle of Bunker Hill took place. It was completed in 1843. Stairs give access to the top of the obelisk affording a great view of Boston. Across the street from the monument is the Bunker Hill Museum, which is home to exhibits and artifacts of the battle, as well as to a massive cyclorama depicting the events of the engagement.

---

6. USS CONSTITUTION

1 Constitution Road, Charlestown, Massachusetts, 02129

Site Type: Wooden Frigate
Conflicts: Barbary Wars, War of 1812
Dates: Commissioned in 1797
Designations: National Register of Historic Places, National Historic Landmark
Web: www.navy.mil/uss-constitution (official website)

The USS Constitution, a colonial era frigate, is America's most famous pre-ironclad sailing ship. Completed in 1797, it was one of the earliest vessels commissioned by the United States Navy. It was also one of America's most accomplished warships, earning the nickname *Old Ironsides* during the War of 1812. It continued to see use throughout the 19th century, for naval training, ambassadorial missions and goodwill tours. In the early 1900s it was re-designated for use as a museum ship.

The Constitution was the third of the original six ships ordered by the United States Navy in the early days of the Republic. Within a few years of its completion it was sent into active wartime service against the Barbary pirates in North Africa, America's first overseas conflict. The Constitution saw its first action at the Battle of Tripoli Harbor in 1803.

Less than a decade later the Constitution earned the admiration of the American public during the War of 1812. She defeated a number of British ships of the line during the course of the war, remaining unbeaten during the conflict. After the War of 1812 the Constitution performed an assortment of duties, including goodwill tours and as a training vessel for naval cadets. The Constitution was added to the National Register of Historic Places in 1966.

The USS Constitution boasts numerous records and accomplishments. It is the oldest commissioned vessel in the United States Navy, and one of the oldest commissioned warships anywhere in the world. It is also one of the oldest ships afloat that can still sail under its own power. The Constitution has an honorary crew which takes care of the ship and keeps it open for tourism. It is now permanently moored at the Charlestown Navy Yard in Boston and is part of Boston's Freedom Trail. Next door is the USS Constitution Museum, an independent institution with exhibits and artifacts from the ship's history.

---

7. MASSACHUSETTS 54TH REGIMENT MEMORIAL

Beacon Street & Park Street, Boston, Massachusetts, 02108

Site Type: Monument – Massachusetts 54th Regiment
Conflict: American Civil War
Dates: Dedicated in 1897
Web: www.boston.gov/parks/boston-common (official website of Boston Common)

The Massachusetts 54th Regiment Memorial, full name the *Memorial to Robert Gould Shaw and the Massachusetts 54th Regiment*, honors the formation and accomplishments of the first all-black regiment to serve in the United States Army. The 54th was formed in 1863 under the command of Robert Shaw and was active until the end of the war. They are perhaps most famous for their assault on Fort Wagner, as depicted in the 1989 film *Glory*.

Troops of African descent had fought in a number of America's earlier wars, including the Revolution and the War of 1812. But it wasn't until the Civil War that African American troops were formally recruited into the army in large numbers. The first all-black units began forming in 1863, and by the end of the war in 1865 approximately one-hundred and eighty thousand African Americans were serving in uniform. Of those who served, none were more famous than the men of the Massachusetts 54th.

Formed in early 1863, the 54th entered service in May, and by July were deployed to South Carolina. They fought in several minor engagements early in the month, and were instrumental in stopping a Confederate advance at the Battle of Grimball's Landing. Their most famous moment came on July 18 at the Battle of Fort Wagner. The 54th led the main assault on the fort, nearly carrying the day. Although they were eventually driven back with heavy losses, word of their bravery and heroics reached Washington and encouraged Lincoln to increase the army's recruitment efforts among African Americans.

The Memorial to Robert Gould Shaw and the Massachusetts 54th Regiment was dedicated in 1897. It depicts Colonel Shaw on horseback riding alongside soldiers of the regiment on the march. Originally intended to be set up in South Carolina, it was instead erected in Boston where the regiment was formed. It is now on display on the Boston Commons.

---

8. MOUNT AUBURN CEMETERY

580 Mount Auburn Street, Cambridge, Massachusetts, 02138

Site Type: Cemetery
Dates: Opened in 1831
Web: http://mountauburn.org (official website)

The Mount Auburn Cemetery is home to one of America's more unlikely gravesites of military interest: that of Julia Ward Howe. Howe was one of America's great cultural figures of the 19th century, known for her writings and her causes, in particular the Abolitionist movement. But she is perhaps most famous for writing one of the greatest American wartime songs, *The Battle Hymn of the Republic*.

Julia Ward Howe was born into a prominent New York City family in 1819. She spent most of her adult life pursuing the advancement of social justice including the abolition of slavery. In 1861 she and her husband visited the Lincolns at the White House, where she was inspired to write the lyrics to *The Battle Hymn of the Republic*, set to the then popular tune of *John Brown's Body*.

The lyrics were subsequently published in Harper's Weekly. The song was immediately popular and became the de-facto anthem of the Union army throughout the Civil War. It remains a popular wartime tune, as well as a church hymn, to the present day. In 1970 Howe was inducted into the Songwriter's Hall of Fame. She died in 1910 and was buried in the Mount Auburn Cemetery.

Mount Auburn Cemetery is an historic burial ground in Boston, with two century's worth of Boston's leading citizens interred here. In addition to Julia Ward Howe, there are a number of prominent military burials, including General Charles Lowell and Colonel Charles Walcott, both of whom fought in the American Civil War.

9. MINUTE MAN NATIONAL HISTORICAL PARK

174 Liberty Street, Concord, Massachusetts, 01742

Site Type: Battlefield - Battle of Lexington and Concord
Conflict: American Revolution
Dates: Battle fought on April 19, 1775
Designations: National Historical Park
Web: www.nps.gov/mima (official website)

The Battle(s) of Lexington and Concord was the first true military engagement of the American Revolution. Immortalized by Henry Wadsworth Longfellow's poem *Paul Revere's Ride*, Lexington and Concord pitted several thousand American militiamen against a force of highly professional and well armed British regulars in a series of engagements outside of Boston in April 1775.

The day of the battle began with approximately seven hundred British soldiers marching from Boston intent on dealing with rebel activity in the countryside. The American militia was warned of the British activity by the lighting of lanterns in the bell tower of the Old North Church in Boston and the riding of messengers to call the local Minute Men to arms. By the time the British reached the town of Lexington, several hundred American militiamen were waiting for them.

Following a brief first skirmish, the Americans retreated with the British in hot pursuit. However, as the day drew on, more and more militiamen joined the American force, and by the time the two sides reached Concord, the rebels were prepared. Fighting led to heavy casualties on both sides, but by afternoon the British were forced to abandon the field. The return to Boston nearly became a rout. The battle ended with the British besieged in Boston. Minute Man National Historical Park was established in 1959.

Minute Man National Historical Park incorporates a number of sites spread out along Battle Road Trail where the day-long engagement took place. A portion of the route between Lexington and Concord has been restored to look as it did during the Colonial era, including the locally famous Hartwell Tavern. Other highlights of the battlefield include the North Bridge at Concord, where the Massachusetts militia rallied to stop the British advance, and the Paul Revere Monument, marking the place where the famous silversmith was captured.

---

10. USS SALEM & UNITED STATES NAVAL SHIPBUILDING MUSEUM

549 South Street, Quincy, Massachusetts, 02169 (relocating at the time of this writing)

Site Type: Heavy Cruiser
Conflict: Cold War
Dates: Commissioned in 1949
Web: www.uss-salem.org (official website)

The USS Salem is the only surviving heavy cruiser of the Des Moines class. Begun during World War II, it was among the very last capital warships, excluding aircraft carriers, to be completed, and one of the last to see service. Active primarily in the Atlantic and Mediterranean during the post-war era, the Salem saw little military action throughout its relatively short ten year career. It was re-opened as a museum ship in Quincy in the 1990s.

The Salem entered active service in 1949, spending most of its time in the Mediterranean with the American 6th fleet. She served mostly as a presence of deterrence in the Eastern Mediterranean during several crises, notably the Suez War. The Salem stood in for the German battleship Graf Spee during the filming of the 1956 movie *The Battle of the River Platte*. She was decommissioned in 1959.

The USS Salem is the only heavy cruiser currently in use as a museum ship. Home to the United States Naval Shipbuilding Museum which opened in 1993, it includes exhibits on several World War II era ships as well as the Navy Seals. Note - As of the time of this writing, the USS Salem was in the process of being relocated to the Fore River Shipyard.

---

11. USS MASSACHUSETTS & BATTLESHIP COVE MARITIME MUSEUM

5 Water Street, Falls River, Massachusetts, 02721

Site Type: Battleship
Conflict: World War II
Dates: Commissioned in 1942
Designations: National Register of Historic Places, National Historic Landmark
Web: https://battleshipcove.org (official website)

The USS Massachusetts (BB-59), or *Big Mamie*, was a South Dakota class battleship that was completed during World War II. It was one of the only American battleships that saw action in both the Atlantic and Pacific theaters, and was one of only a handful of American ships that fought directly against the naval forces of Vichy France, an ally of Nazi Germany. The Massachusetts' active service was short, less than five years, before she was decommissioned in 1947. It was established as a museum ship at Battleship Cove in 1965.

The Massachusetts entered active duty in the Atlantic theater of World War II, fighting its first major engagement less than six months after being commissioned. Part of the escort fleet that supported the American invasion of North Africa in November 1942, the Massachusetts fought at the battle of Casablanca, disabling one French battleship and contributing to the sinking of a cruiser and two destroyers. This action paved the way for the Allied liberation of Morocco in the following weeks.

After North Africa the Massachusetts sailed to the Pacific, where it was involved in many important naval actions against Japan. Among these were the campaigns in the Solomon Islands, the Caroline Islands, Leyte Gulf and Iwo Jima. The Massachusetts also participated in several bombardments of the Japanese mainland before the war came to a close. The USS Massachusetts was added to the National Register of Historic Places in 1976 and named a National Historic Landmark in 1986.

The USS Massachusetts is now located at Falls River where it is part of the Battleship Cover Maritime Museum. In addition to the historic battleship, visitors can explore the remains of the USS Joseph P. Kennedy Jr., the submarine USS Lionfish, and several PT boats. Also here is the Hiddensee, an East German vessel acquired after the end of the Cold War, and a piece of the bow of the USS Falls River, a former cruiser named for the city.

---

12. SPRINGFIELD ARMORY NATIONAL HISTORIC SITE

1 Armory Square, Springfield, Massachusetts, 01105

Site Type: Arsenal
Conflict: Shay's Rebellion
Dates: Opened in 1778
Designations: National Register of Historic Places, National Historic Site, Springfield National Historic Landmark District
Web: www.nps.gov/spar (official website)

The Springfield Armory was one of the first and for many years among the most important military arsenals in the United States. Tracing its roots back to the American Revolution, the Springfield Armory remained in active use for nearly two centuries. It remained open all the way through World War II, but was obsolete by the time of the Cold War and closed in the 1960s.

In 1777 George Washington, at the prompting of Henry Knox, decided Springfield would be a strategically advantageous site for a military arsenal to support the rebellion. Already in use by local militia, a permanent armory was completed in 1778. Used throughout the American Revolution, the facility was kept open for continued use after the war.

The armory played a small but direct role in Massachusetts history in 1787, when partisans of Shays' Rebellion marched on Springfield in order to secure its store of weapons. A skirmish took place just outside of the armory, and Shays' men were forced to withdraw. The rebellion was crushed shortly thereafter. The armory remained in active use until 1968. It was included in the Springfield National Historic Landmark District in 1960 and listed on the National Register of Historic Places in 1966.

The Springfield Armory was thoroughly renovated after its closure and is now run as a museum. It is home to one of the largest collections of firearms to be found in the United States. Racks of guns from every era of American history fill the museum, recalling the days when hundreds of thousands of firearms were stored here for wartime use. Highlights of the collection include exhibits on small arms and machine guns.

---

13. PATRIOT'S PARK

RI 114, Portsmouth, Rhode Island, 02871

Site Type: Battlefield – Battle of Rhode Island
Conflict: American Revolution
Dates: Battle fought on August 29, 1778
Designations: National Register of Historic Places, National Historic Landmark
Web: www.portsmouthri375.com/patriots-park (official website)

The Battle of Rhode Island was one of the last major engagements of the American Revolution to take place in New England. Considered a tactical draw with a minor advantage for the British, the battle was perhaps most famous due to the fact that it was here for the first time that soldiers of African ancestry fought for the American army. The 1st Rhode Island Regiment, more than half of which was composed of black soldiers, has been honored for playing a distinguished role in the battle.

In the aftermath of their defeat at Saratoga and the subsequent entry of France into the war on the American side, the British strengthened their position in the city of Newport, one of the key New England ports still in British hands. In the summer of 1778 the Americans and French decided to make the retaking of Newport their first major joint operation. A large force set out to lay siege to the city with expectations of support from the French fleet. However, an unexpected storm forced the French ships to withdraw.

Without French naval aid the Americans were forced to abandon the campaign and fall back. They were pursued by a British army from Newport and the two sides met on August 29. The Americans held off several British assaults, with the 1st Rhode Island Regiment playing a crucial role. They were later praised by French commander Lafayette for their actions. Although the Americans were forced to retreat, they withdrew in good order, keeping the army intact. Patriot's Park was added to the National Register of Historic Places and named a National Historic Landmark in 1974.

Patriot's Park incorporates much of the area in which the engagement took place, including Butts Hill and Turkey Hill. The former still has intact earthworks which were used by the Americans at the battle. A large stone monument honors those from Rhode Island who were killed in the engagement. There is also a monument specifically commemorating the contribution of the African American soldiers who fought here.

---

14. GREAT SWAMP FIGHT MONUMENT

41 Great Swamp Monument Road, West Kingston, Rhode Island, 02892

Site Type: Battlefield – Great Swamp Fight/Massacre
Conflict: King Philip's War
Dates: Battle fought on December 19, 1675;
Monument erected in 1906
Web: www.southcountyri.com/listing/great-swamp-monument/170 (official county tourism website)

The Great Swamp Fight Monument commemorates the battle, or massacre, that took place in 1675 at a fort that once stood here. The Great Swamp Fight was the largest and one of the most important engagements of the conflict known as King Philip's War, an uprising of Native American tribes throughout New England against the growing number of English colonists. The war raged for three years, and was the first and arguably the greatest effort ever made by Native Americans to dislodge European settlers from the continent.

In 1675 Metacomet, a chief of the Wampanoags, led an alliance of tribes in a desperate effort to stop the spread of English settlements in the region. The ensuing war was one of the bloodiest in early American history, with thousands killed on both sides. While the Native Americans enjoyed some early successes, they were badly defeated at the largest battle of the conflict, the Great Swamp Fight.

During this engagement, colonial militia assaulted the main Native American fort, which had been constructed in a swamp as a defensive measure. However, during the winter, the swamp froze over, making it easy for the militia to attack over the ice. The ensuing battle was fierce, with over two hundred militia killed or wounded. However, Metacomet's force of about a hundred warriors was killed, with perhaps as many as a thousand non-combatants killed or captured.

The Great Swamp Fight Monument consists of a rough-hewn obelisk and several stone markers which stand on what is believed to have been the site of the battle. While it primarily commemorates the colonial victory and the militia who fought and died here, the marker also specifically acknowledges the valiant stand of the local tribes. The monument is now regarded by many as a reminder of the massacre of the native peoples of New England and the beginning of a futile, centuries-long struggle by Native Americans to defend their lands.

15. UNITED STATES COAST GUARD ACADEMY

31 Mohegan Avenue, New London, Connecticut, 06320

Site Type: Military Academy
Dates: Founded in 1930
Web: www.uscga.edu (official website)

The United States Coast Guard Academy is a maritime college for the training of American coast guard officers. Founded in 1876 as the School of Instruction of the Revenue Cutter Service, it is the third oldest of the federal service academies as well as the smallest. Nevertheless it has a long history of tradition and contribution to the activities of the American military at sea.

The Coast Guard Academy began as a training school sponsored by the United States Treasury Department as a means to enforce customs. In 1930 the school was re-designated as the United States Coast Guard Academy and greatly expanded. The academy has occupied the present site and has been in operation since 1932.

The United States Coast Guard Academy campus is relatively small compared to West Point and Annapolis, but there are some interesting things worth seeing. One of these is the United States Coast Guard Museum, which has exhibits on the history of the Coast Guard that include everything from model ships to uniforms to cannons. Also on site is the USCGC Eagle, one of only two sailing vessels that are still in active American military service. Seized from Nazi Germany after World War II, it is now used for cadet training.

16. FORT TRUMBULL STATE PARK

90 Walbach Street, New London, Connecticut, 06320

Site Type: Masonry Star Fort
Conflict: American Revolution
Dates: Originally completed in 1852
Designations: National Register of Historic Places
Web: www.thamesriverheritagepark.org/fort-trumbull (official website)

Fort Trumbull was one of the longest active coastal fortifications in American history. Originally built in the 1770s, Fort Trumbull was still in military use in the 1990s over two centuries later. It saw action during the American Revolution, served for a time as the home of the United States Coast Guard Academy, and was even used for military research in its final active decades.

The original fort along the Thames River was built in 1777, just in time for service in the Revolution. Garrisoned by members of the Connecticut militia, it was seized by a British force under Benedict Arnold in 1781. After the Revolution the original structure remained in service for several decades. The fort as it currently stands was largely rebuilt in the 1840s.

Fort Trumbull was in active use throughout the 19th century. From 1910 through 1932 the United States Coast Guard Academy was based here before moving to its current location. It continued its role as an educational facility during and after World War II, primarily as a school for the Merchant Marines. In its final years Fort Trumbull was used as a center for underwater military research. The fort was finally decommissioned in 1996. Fort Trumbull was listed on the National Register of Historic Places in 1972.

Fort Trumbull is in very good condition, thanks in large part to its many and mostly continual years of active use. Surrounded by the greenery of a state park, Trumbull looks much as it did at the time of its construction. Its long main battlement, which once accommodated dozens of cannon, offers a magnificent view of the river. The fort's barracks are now home to a museum.

NEW YORK

17. CASTLE CLINTON NATIONAL MONUMENT & BATTERY PARK MEMORIALS

Battery Park, New York, New York, 10004

Site Type: Masonry Fort, Monuments – World War II, Korean War
Dates: Fort originally completed in 1811; Monuments dedicated in 1963 & 1991
Designations: National Register of Historic Places, National Monument, New York City Landmark
Web: www.nps.gov/cacl (official website)

Castle Clinton is the current incarnation of a gun battery installation that once guarded the harbor of New York City. Informally known as The Battery, for which surrounding Battery Park is named, the fort has served in many capacities over the course of two centuries. Battery Park, which covers the southwest tip of Manhattan Island, is also home to a number of major war memorials.

The area now known as the Battery was the site of military fortifications as early as the 17th century. The Dutch constructed Fort Amsterdam here in the 1620s, and it survived under various names until the 1780s when it was annexed by the United States. The obsolete fort was destroyed in 1790 and replaced with a gun battery in 1811. The installation was only active for a few years, and though manned during the War of 1812 never saw military use.

The fort was decommissioned in 1821 and taken over by the city of New York. It subsequently served as, among other things, a theater, an immigration checkpoint and an aquarium. It is currently being used as a transportation center for ferries to the Statue of Liberty and Ellis Island. Castle Clinton was designated a National Monument in 1946, named a New York City Landmark in 1965 and listed on the National Register of Historic Places in 1966.

Castle Clinton is a round, red brick fort with apertures for cannons facing the harbor. It stands a short distance from the water's edge, with its back and sides surrounded by Battery Park. During the 20th century Battery Park became a popular place to erect war memorials, especially for those who served and died at sea. Among the monuments here are the East Coast Memorial to the Missing, which honors the thousands of naval servicemen who died during World War II; the Merchant Mariner's Memorial and Coast Guard Memorial, which also honors members of those services who died in World War II; and a Korean War Memorial.

---

18. GEORGE M. COHAN MONUMENT

Times Square, New York, New York, 10036

Site Type: Monument – George M. Cohan
Dates: Dedicated in 1959
Web: www.nycgovparks.org/parks/father-duffy-square (official website of Duffy Square)

The George M. Cohan Monument is one of the best known statues in New York City, even though many people in the 21st century are unfamiliar with this great American composer. Standing in the very center of Times Square, hundreds of thousands of passers-by see the statue every day. Best known for his work on Broadway, Cohan also composed some of the most popular wartime songs ever written, for which he was later awarded a Congressional Medal of Honor.

Cohan was born into a show business family and was raised working in vaudeville. He made a name for himself as a writer, producer and actor on Broadway. His first major hit show, *Little Johnny Jones*, featured the song *The Yankee Doodle Boy*. This established him as a major composer of American patriotic music.

Among his later works were the marches *Grand Old Flag* and *Over There*, which became popular songs for the American Expeditionary Force fighting in Europe during World War I. In recognition of his patriotic efforts Cohan was awarded the Congressional Medal of Honor in 1936. He died in 1942 and was buried in Woodlawn Cemetery in the Bronx.

The Cohan Monument is one of the definitive landmarks of Times Square. It was erected here due to the composer's longtime association with Broadway. The monument is technically located in the adjoining Duffy Square, which is named for Father Francis Duffy, a highly decorated World War I chaplain. Duffy is also honored with a bronze statue here.

---

19. USS INTREPID SEA, AIR AND SPACE MUSEUM

Pier 86, 12th Avenue & West 46th Street, New York City, New York, 10036

Site Type: Aircraft Carrier
Conflicts: World War II, Vietnam War, Cold War
Dates: Commissioned in 1943
Web: www.intrepidmuseum.org (official website)

The USS Intrepid (CV-11), also known as the *Fighting I*, is arguably the most visited aircraft carrier still afloat. Laid down in 1941, less than a week before the Pearl Harbor attack, the Intrepid was completed in time to participate in the war in the Pacific. She remained in service for more than three decades, and was deployed during the Vietnam War before being decommissioned in 1974. In 1982 the carrier was permanently moored on the west side of Manhattan and repurposed as an air and space museum.

The Intrepid was an active combat vessel used primarily during World War II. It saw nearly continuous action from its arrival in the Pacific in early 1944 until the end of the war in 1945. She participated in some of the war's most famous actions, most notably the Battle of Leyte Gulf. The Intrepid was also instrumental in supporting the American invasion of Okinawa in 1945.

After World War II the Intrepid was decommissioned for a few years before returning to service in 1955. It remained on active duty throughout much of the Cold War, including in Vietnam in the late 1960s. The carrier was also used as a recovery vessel for NASA's Mercury and Gemini programs. Although it has served as a museum since 1982 the Intrepid was pressed into service once more in 2001, when it briefly served as a field headquarters for the FBI after the World Trade Center attack.

The Intrepid Sea, Air and Space Museum is one of the great museums of the city of New York. Permanently moored in the Hudson River in Midtown Manhattan, it now exhibits a magnificent collection of both aircraft and spacecraft, including the Space Shuttle Enterprise. There are also exhibits on the Intrepid and her service during World War II. Moored next to the Intrepid is the USS Growler, a decommissioned Cold War era guided missile submarine.

---

20. GENERAL GRANT NATIONAL MEMORIAL

Riverside Drive & West 122nd Street, New York, New York, 10027

Site Type: Monument – Ulysses S. Grant
Dates: Dedicated in 1897
Designations: National Register of Historic Places, National Memorial, New York City Landmark
Web: www.nps.gov/gegr (official website)

The General Grant National Memorial, better known as Grant's Tomb, is an immense monument dedicated to the 18th President of the United States and hero of the American Civil War, Ulysses S. Grant. This hulking mausoleum is one of the great landmarks of New York City's Upper West Side. President Grant and his wife are entombed inside the memorial, despite a long running joke dating from the 1950s suggesting that they are not.

Ulysses S. Grant is ranked among the greatest military leaders in the history of the United States. A graduate of West Point, he served in the Mexican-American War and, most famously, in the American

Civil War. Grant distinguished himself in numerous battles and campaigns, and in 1864 was chosen by President Lincoln to serve as the supreme commander of the army. He personally presided over Robert E. Lee's surrender at Appomattox in 1865.

After the war Grant continued his service in the army, and was formally promoted to General of the Army in 1866, one of only a handful of men in American history to be so honored. In 1868 he was elected President of the United States which he served for two terms. Grant died in 1885 and his monumental tomb completed in 1897. Grant's Tomb was named a National Memorial in 1958, listed on the National Register of Historic Places in 1966 and designated a New York City Landmark in 1975.

The General Grant National Memorial is an imposing neo-classical structure that overlooks the Hudson River at the northern end of Riverside Park. The magnificent interior of the monument is dominated by the sarcophagi of Grant and his wife, Julia. Murals high on the walls depict scenes from Grant's service in the army. Busts of several generals who served with Grant are displayed in niches around the monument.

---

21. WHITE PLAINS BATTLEFIELD SITES

60 Park Avenue, White Plains, New York, 10603

Site Type: Battlefield – Battle of White Plains
Conflict: American Revolution
Dates: Battle fought on October 28, 1776
Designations: National Register of Historic Places
Web: www.visitwestchesterny.com/listing/jacob-purdy-house/67 (official website of Purdy House)

The Battle of White Plains was the site of a British victory that effectively secured Royalist control of New York City for most of the American Revolution. It was one of a series of British victories that saw the American Continental Army chased from Long Island to

Pennsylvania throughout the second half of 1776. It was also a missed opportunity for the British to destroy Washington's army in the field, and therefore can be seen as part of a long, successful delaying action on the part of the Americans.

In the late summer of 1776 the primary British army in the Northern colonies, under the command of William Howe, began a campaign to crush the American rebellion in and around New York City. Between August and October the Americans, under the command of George Washington, were driven out of Long Island and then Manhattan. Howe pursued Washington's shrinking force northwards into Westchester where the two main armies met again at White Plains.

The fighting was fierce, but eventually the superior numbers of the British forced the Americans out of their positions. Despite the defeat Washington's force retreated in reasonably good order, and most of the American army escaped across the Hudson River. The last colonial troops in New York were driven out of Fort Washington in Manhattan a few weeks later. From the 1930s to the 1950s the White Plains battlefield was recognized as a National Battlefield Site, but has since been delisted. The Jacob Purdy House is listed on the National Register of Historic Places.

The White Plains Battlefield is no longer an officially organized park, but the city of White Plains maintains several monuments and buildings related to the engagement. There is the Battle of White Plains historic site marker, a remnant of its National Park Service days. There are two colonial era buildings of interest: the Jacob Purdy House and the Elijah Miller House, both of which were used by George Washington during the war. The former is now home to the White Plains Historical Society, while the latter houses a museum.

22. UNITED STATES MILITARY ACADEMY

606 Thayer Road, West Point, New York, 10996

Site Type: Military Academy
Dates: Founded in 1802
Designations: National Register of Historic Places, National Historic Landmark
Web: www.westpoint.edu (official website)

The United States Military Academy, commonly referred to as West Point, is the primary college for the training of American army officers. Founded in 1802 it is America's oldest professional military school, and counts among its graduates some of the most famous figures in American history. Although most of the campus is not open to visitors, especially when school is in session, there are a number of historic sites at the Academy that can be visited by tour.

The site where the United States Military Academy now stands was home to fortifications going back to before the beginning of the American Revolution. West Point was seized by Colonial troops in 1778, and has been an American army installation ever since. The fort, which guarded the Hudson River between New York City and Albany, was nearly lost due to the treacherous actions of Benedict Arnold. However, Arnold's plot came unraveled, and West Point remained in rebel hands.

After the war West Point was used as an army training center. It was formally organized into the United States Military Academy in 1802 by act of Congress. Officers who graduated from West Point began to rise to military prominence during the Mexican-American War, and formed the core of America's higher ranking army officers in every war after that. Notable alumni include Robert E. Lee, John J. Pershing, Douglas MacArthur, George Patton, and presidents Ulysses S. Grant and Dwight D. Eisenhower.

The United States Military Academy is open on a limited basis due to both security and the fact that this is an active campus. Tours begin at the visitor's center, which has exhibits on the academy and cadet life. The adjacent West Point Museum has a fascinating and unusual assortment of military artifacts. One of the most popular destinations for visitors is the West Point Cemetery where the gravesites of nearly two centuries of distinguished alumni can be found, including that of George Armstrong Custer.

---

23. BENNINGTON BATTLEFIELD STATE HISTORIC SITE

5157 Route 67, Walloomsac, New York, 12090

Site Type: Battlefield – Battle of Bennington
Conflict: American Revolution
Dates: Battle fought on August 16, 1777
Designations: National Register of Historic Places, National Historic Landmark, New York State Historic Site
Web: https://parks.ny.gov/historic-sites/12/details (official website)

The Battle of Bennington was named for Bennington, Vermont, the objective of the British campaign. But the actual battle took place ten miles to the west in what is now New York State. A resounding victory for the Americans, the loss at Bennington was a major contributing factor to the even greater British loss at Saratoga a few weeks later. The battle is commemorated both in New York where the fighting took place and at Bennington in Vermont.

Despite impressive victories in New York and New Jersey in 1776, the British were having a difficult time keeping rebellious New England under control. So the next year they decided to focus on isolating New England from the other colonies by taking control of the Hudson River Valley. This campaign, organized by John Burgoyne, proved slow going, and by August the British were running low on supplies. To alleviate this situation the British sent a foraging expedition to Bennington in order to secure food and munitions.

Expecting to find Bennington guarded by a token militia force, they found themselves outnumbered by a recently raised army from

New Hampshire under the command of John Stark. The two sides met at Walloomsac west of Bennington, where a well executed flanking maneuver left the British completely surrounded. The result was a decisive victory for the Americans, with nearly two thirds of the enemy killed or captured, crippling Burgoyne's main force. The Bennington Battlefield State Historic Site was named a National Historic Site in 1961 and listed on the National Register of Historic Places in 1966.

The Bennington Battlefield State Historic Site incorporates the bulk of the areas where the battle took place. There are numerous stone monuments around the park noting where events of the engagement occurred, and memorials to the New Hampshire, Vermont and Massachusetts militiamen who fought here. The main monument commemorating the battle is located in nearby Bennington, Vermont.

---

24. FORT WILLIAM HENRY MUSEUM

48 Canada Street, Lake George, New York, 12845

Site Type: Wooden Star Fort
Conflict: French and Indian War
Dates: Originally completed in 1755
Web: www.fwhmuseum.com (official website)

Fort William Henry and the nearby area of Lake George was the location of some of the fiercest fighting during the French and Indian War. The siege of the fort and the subsequent massacre of the British who surrendered were popularized by James Cooper in his immortal novel about life in colonial New York, *The Last of the Mohicans*. The current fort is largely a recreation and is now operated as a living museum.

The first fighting in the area around the fort took place in 1755 at the Battle of Lake George. This early engagement saw several skirmishes which resulted in a British victory. In order to better secure the area against further French attacks, the British built Fort William Henry. They subsequently held control of the area for two years before the French arrived with a much larger force intent on driving them out.

In the spring and summer of 1757 a French army in the area forced many locals to seek shelter at Fort William Henry. In August the French arrived at the fort and laid siege to it. After a week, when it became clear that they would not be able to hold out without reinforcements, the British surrendered. Theoretically under the protection of the French, the fort inhabitants marched to the safety of nearby Fort Edward. However, they were harassed en-route by Huron warriors allied to the French, and many of the British were killed along the march. After taking the fort, the French razed and abandoned the site.

Fort William Henry was left in ruins for the better part of two centuries before being partially reconstructed in the 1950s. The site incorporates some of the original earthworks along with part of the outer walls. The fort is run as a living museum, with employees in period costume offering talks and military demonstrations on certain days. Also nearby is the Lake George National Battlefield Park, an archaeological site where fighting took place during the French and Indian War.

---

25. FORT TICONDEROGA NATIONAL HISTORIC LANDMARK

102 Fort Ticonderoga Road, Ticonderoga, New York, 12883

Site Type: Masonry Star Fort
Conflicts: French and Indian War, American Revolution
Dates: Originally completed in 1758
Designations: National Register of Historic Places, National Historic Landmark
Web: www.fortticonderoga.org (official website)

Fort Ticonderoga in upstate New York was originally built by the French to guard the approach to Canada against the British in New York. It was the site of a number of major engagements during the French and Indian War and the American Revolution. After the Revolution the fort largely fell into a state of disrepair. Much of Fort

Ticonderoga has been restored in stages over the course of the last century.

During the 1750s, the Seven Years War in Europe spilled over to the North American continent in the form of the French and Indian War. To defend against a British invasion up the Hudson River Valley, the French constructed Fort Carillon at the southern end of Lake Champlain. It was completed in 1758, just in time to defend against and stop a major British attack. The British returned the next year and seized the fort from a greatly reduced French garrison.

In May 1775, just a few weeks after the American Revolution began, militia from Massachusetts and Connecticut under the command of Benedict Arnold seized the fort. It briefly served as the base of American operations for their aborted invasion of Canada. Fort Ticonderoga was retaken by the British in 1777 and finally abandoned in 1781 at the end of the war. The fort spent most of the 19th century as a partial ruin before restoration efforts began in the early 20th century. Fort Ticonderoga was designated a National Historic Landmark in 1960 and added to the National Register of Historic Places in 1966.

Fort Ticonderoga consists partially of original structures and partially of reconstructed walls and buildings. The ramparts and bastions look much as they did in the 18th century and feature period cannons on display. Several barracks buildings as well as the reconstructed powder magazine and stores buildings are now home to a museum with exhibits on the history of the fort.

---

26. SARATOGA NATIONAL HISTORICAL PARK

648 Route 32, Stillwater, New York, 12170

Site Type: Battlefield - Battle of Saratoga
Conflict: American Revolution
Dates: Battles fought on September 19, 1777 & October 7, 1777
Designations: National Historical Park
Web: www.nps.gov/sara (official website)

The Battle of Saratoga was a series of two major and several minor engagements in Upstate New York which effectively turned the tide of the American Revolution in the northern colonies. Some of the war's most famous commanders from both sides were present at the battle, including Horatio Gates, Daniel Morgan and a pre-traitor Benedict Arnold leading the Americans. The battle effectively ended British efforts to win the war in New England and New York.

After the failed American invasion of Canada in 1775, the British sought to secure the Hudson River Valley in order to divide New England from the rest of the colonies. A British force under John Burgoyne came down from Canada in the summer of 1777, recapturing Fort Ticonderoga before proceeding towards Albany. On September 19 they met fierce American resistance outside of Saratoga. Though the British drove the Americans from the field, the battle was indecisive.

A few weeks later, a strongly reinforced American army counter-attacked a reduced British force at Bemis Heights on October 7. The result was an overwhelming American victory, one of the greatest of the war. After Saratoga the British had completely lost the initiative in the northern colonies and were forced to garrison defensive positions there for the rest of the war. The American success also encouraged the French to join the war effort on the American side. Saratoga National Historical Park was established in 1938.

Saratoga National Historical Park includes a large wooded area where much of the battle took place on October 7. On the property is the restored home of American Revolution general and hero Philip Schuyler. The centerpiece of the park is a huge obelisk which marks the site of Burgoyne's surrender. American commanders Gates, Morgan and Schuyler are honored, while Benedict Arnold is conspicuously absent. Arnold is commemorated by a small monument which does not technically mention him by name.

---

27. FORT STANWIX NATIONAL MONUMENT

100 North James Street, Rome, New York, 13440

Site Type: Wooden Star Fort
Conflicts: French and Indian War, American Revolution
Dates: Originally completed in 1762
Designations: National Register of Historic Places, National Historic Landmark, National Monument
Web: www.nps.gov/fost (official website)

Fort Stanwix was constructed by the British during the French and Indian War. However, it primarily saw action during the American Revolution. Seized by Colonial troops in the early days of the war, the British attempted unsuccessfully to retake it. Failure to secure the fort, combined with their defeat at the Battle of Bennington, directly contributed to the overwhelming American victory at Saratoga a few months later.

During the French and Indian War the British constructed Fort Stanwix to guard against attack down the Mohawk River Valley. It was largely abandoned in the mid-1760s, though a peace negotiation was held here between the British and Iroquois in 1768. In July 1776 an American force occupied the site of the fort and rebuilt it, renaming it Fort Schuyler.

A year later, during their campaign to secure Albany and the Hudson River Valley, the British sent a force westward to retake the fort. After a brief siege in August of 1777 they abandoned the effort. The Americans held the fort until 1781 when it was burned down and abandoned. Fort Stanwix was designated a National Monument in 1935, named a National Historic Landmark in 1962 and listed on the National Register of Historic Places in 1966.

Fort Stanwix National Monument is for the most part a replica of the original fort that once stood here in the 18th century. It was completed in the 1970s after an extensive archaeological survey of the area. A traditional colonial era gunpowder star fort, it is unusual in that the walls are made out of wood rather than stone. It nevertheless has the style of earthworks more common to the stone forts of the period.

28. BUFFALO AND ERIE COUNTY NAVAL AND MILITARY PARK

1 Naval Park Cove, Buffalo, New York, 14202

Site Type: Museum
Dates: Opened in 1979
Web: http://buffalonavalpark.org (official website)

The Buffalo and Erie County Naval and Military Park is home to a collection of military vehicles, aircraft and warships, and is one of the few such museums in which land, sea and air vehicles are represented. Many machines of war can be found here, from tanks to helicopters to fighter jets, though the centerpiece attractions are the destroyer USS Sullivans and the light cruiser USS Little Rock.

The USS Little Rock entered active military service in 1945 but not in time to serve in World War II. After several inactive years the Little Rock was refitted as a guided missile cruiser in 1957. She was active in the Caribbean, Atlantic and Mediterranean before being finally decommissioned in 1976. The USS Sullivans was commissioned in 1943 and was directly involved in many battles in the Pacific Theater. The Sullivans also saw action during the Korean War. She was finally decommissioned in 1965.

The Buffalo and Erie County Naval and Military Park is a very diverse attraction as military museums go. The Little Rock and the Sullivans were the initial naval exhibits at the Buffalo and Erie County Naval and Military Park. They were later joined by a submarine and PT boat. Military vehicles on display at the museum are an M-84 armored personnel carrier and an M-41 tank. There are also several aircraft on display, including a UH-1 helicopter, an FJ-4B fighter jet, and a World War II era P-39 fighter.

NEW JERSEY & PENNSYLVANIA

29. MORRISTOWN NATIONAL HISTORICAL PARK

30 Washington Place, Morristown, New Jersey, 07960

Site Type: Encampment
Conflict: American Revolution
Dates: Camp established in 1779
Designations: National Register of Historic Places, National Historic Park, New Jersey Register of Historic Places
Web: www.nps.gov/morr (official website)

During the winter of 1779-1780, George Washington and the Continental Army made their camp near Morristown, New Jersey. Although this camp was established after the tide of the war in the northern colonies had already turned against the British, the winter, one of the coldest in years, was still a hardship for the Americans. It was here that the rebels began to prepare for the final two years of campaigning that would ultimately see them victorious in 1781.

1779 was one of the quieter years of the American Revolution, with relatively few major engagements as the British prepared to redirect the war to the southern colonies. In the north, an American raid on Paulus Hook effectively gave the rebels a free hand in New Jersey. It was for this reason, as well as its excellent strategic location, that Washington chose Morristown for his winter camp.

What should have been an uneventful season turned into one of the coldest winters on record, even worse than at Valley Forge. At one point the Continental Army faced a mutiny by troops from Pennsylvania and New Jersey, but overall the American force survived essentially intact for the 1780 campaigning season. In 1933, Morristown was named America's first National Historical Park. It was listed on the

National Register of Historic Places in 1966 and the New Jersey Register of Historic Places in 1971.

Morristown National Historic Park was the site of the 1779 winter encampment and features four primary sites of interest. Fort Nonsense, a hilltop which may have been the site of a signal fire, and which may have been constructed as a means of keeping the soldiers occupied; Jockey Hollow, where the army was encamped; Ford Mansion, where Washington lived with his family during the winter; and the Washington Headquarters Museum. A related place of interest, the New Jersey Brigade Encampment, is located nearby.

---

30. MONMOUTH BATTLEFIELD STATE PARK

16 Business Route 33, Manalapan, New Jersey, 07726

Site Type: Battlefield - Battle of Monmouth
Conflict: American Revolution
Dates: Battle fought on June 28, 1778
Designations: National Register of Historic Places, National Historic Landmark District, New Jersey Register of Historic Places
Web: www.state.nj.us/dep/parksandforests/parks/monbat.html (official website)

The Battle of Monmouth was one of the last major engagements of the American Revolution in the northern colonies. Mostly a test of the Continental Army's readiness after their winter at Valley Forge, this American attack on the British army was inconclusive but a major morale boost for the rebels. The Battle of Monmouth followed the British evacuation of Philadelphia and effectively left the English bottled up in New York City for most of the rest of the war.

In 1778 France entered the war on the American side. The sudden appearance of a modern European army and navy in North America forced the British to completely rethink their strategy. This included the abandoning of Philadelphia in favor of strengthening their hold on New York City. After Philadelphia was restored to American control the Continental Army decided to test the enemy by harassing their march to New York.

On June 28 the American army attacked the British rearguard and a major engagement ensued. Attacks were made by both sides in blistering heat that took a heavy toll. By the time the battle ended in what was effectively a draw, each side had taken significant casualties. Nevertheless the Americans were left in control of the field, and the British pressed on to New York. Monmouth Battlefield was named a National Historic Landmark District in 1961, listed on the National Register of Historic Places in 1966, and the New Jersey Register of Historic Places in 1971.

Monmouth Battlefield State Park is considered by enthusiasts as one of the best preserved battle sites in the Northeast. Several houses that existed at the time of the battle are still in existence, and several monuments and markers are located throughout the property. One of these is a statue of Friedrich von Steuben, the German officer who helped to train the Continental Army at Valley Forge. There is also a small memorial to Molly Pritcher who, according to tradition, took her husband's place as an artilleryman at the height of the battle.

---

31. USS NEW JERSEY

62 Battleship Place, Camden, New Jersey, 08103

Site Type: Battleship
Conflicts: World War II, Korean War, Vietnam War, Cold War
Dates: Commissioned in 1943
Designations: National Register of Historic Places, New Jersey Historical Place
Web: www.battleshipnewjersey.org (official website)

The USS New Jersey (BB-62), or the *Big J*, was one of the longest-serving American battleships of the 20th century, and one of the last to be used in a combat capacity. Launched on the one-year anniversary of the Pearl Harbor attack, the New Jersey was one of the chief battleships active in the Pacific theater during World War II. It later saw service in Korea and Vietnam, as well as other activities during the Cold War, before being decommissioned in 1991. It was permanently moored in Camden in 2000 for use as a museum ship.

The New Jersey entered service in World War II in early 1944 as part of the invasion of the Marshall Islands. She served in a number of Pacific campaigns, occasionally as a flagship. Among New Jersey's battles were Leyte Gulf and the invasion of Iwo Jima. After World War II the New Jersey was part of the Atlantic Reserve Fleet.

Over the next few decades the New Jersey was periodically recommissioned, serving in active duty in both the Korean and Vietnam wars, and was part of the peacekeeping force that served in the Mediterranean in the 1980s. By the time of its retirement, it had earned over twenty battle stars and was the most decorated naval vessel in American history. The USS New Jersey was added to the National Register of Historic Places and named a New Jersey Historical Place in 2004.

The Battleship New Jersey is now permanently located in Camden, New Jersey. Fully restored, at least to the point where it can accommodate visitors, it looks much as it did at the time of its last active service. In addition to the many areas of the ship that can be explored, there are also exhibits on the New Jersey's long and historic service.

---

32. TRENTON BATTLE MONUMENT

348 North Warren Street, Trenton, New Jersey, 08625

Site Type: Battlefield – Battle of Trenton
Conflict: American Revolution
Dates: Battle fought on December 26, 1776;
Monument dedicated in 1893
Designations: National Register of Historic Places, New Jersey Register of Historic Places
Web: www.state.nj.us/dep/parksandforests/historic/Trentonbattlemonument/index.htm (official website)

The Battle of Trenton was one of the greatest morale victories in American history. It was also pivotal in keeping the rebel army, and cause, going in the wake of months of military defeats. Closely associated with Washington's storied crossing of the Delaware River, which immediately preceded the battle, many Americans consider this to be one of the most important engagements of the war.

In the second half of 1776 George Washington's Continental Army had been driven from Long Island and New York City across New Jersey and into Pennsylvania in a string of defeats, leaving it in tatters and at a fraction of its original strength. By December it was on the brink of collapse. However, intelligence brought to Washington revealed a potential chink in the British armor: a large Hessian force based in nearby Trenton which was probably going to overindulge in the Christmas holiday and thus unlikely to be prepared for an attack.

In the early morning hours on the day after Christmas the Americans marched on Trenton. The Hessians were caught almost completely unprepared. Despite a desperate resistance, the Hessians were overwhelmed, taking nearly a thousand casualties while the Americans took less than ten. The decisive victory boosted American morale when it was most desperately needed, and held the army together long enough to participate in the Saratoga campaign the next year. The Trenton Battle Monument was listed on the New Jersey Register of Historic Places in 1976 and the National Register of Historic Places in 1977.

The Trenton Battle Monument is an immense victory column crowned with a statue of George Washington. Completed in 1893, it stands on the spot where an American artillery battery was stationed during the battle in order to control the major streets of Trenton throughout the engagement. Several bronze reliefs on the base of the column depict the river crossing.

33. WASHINGTON CROSSING PARKS

*355 Washington Crossing-Pennington Road,
Titusville, New Jersey, 08560*

1112 River Road, Washington Crossing, Pennsylvania, 18977

Site Type: Event – Crossing of the Delaware
Conflict: American Revolution
Dates: Crossing took place on December 26, 1776
Designations: National Register of Historic Places, New Jersey State Park, Pennsylvania Historic Marker
Web: www.state.nj.us/dep/parksandforests/parks/washcros.html & www.washingtoncrossingpark.org (official websites)

Washington Crossing State Park in New Jersey and Washington Crossing Historic Park in Pennsylvania are the two historic sites that commemorate George Washington's crossing of the Delaware River in 1776. These parks can be found at the locations where, on the night of December 25-26, a force of over two thousand men under the command of Washington completed the crossing of the river en route to their surprise attack on the Hessian garrison in the city of Trenton.

After having been driven across New Jersey in a series of campaigns that left the Continental Army battered, Washington decided on a bold but desperate gamble: to attack the unprepared Hessian mercenaries at Trenton on the day after Christmas. On Christmas night they arrived at the Delaware River near Yardley and boarded a flotilla of small boats. Most of Washington's force arrived safely on the other shore near Titusville. A few hours later they pulled off one of the greatest upset victories of the war. Washington Crossing Historic Park was granted a Pennsylvania Historic Marker in 1947 and added to the National Register of Historic Places in 1966.

Washington Crossing Historic Park commemorates the departure point of the operation, and has a number of important sites, including the boathouse where replicas from the crossing are on display. Nearby is the McConkey's Ferry Inn, which some sources indicate is the location where Washington planned the surprise attack. Washington Crossing State Park commemorates the arrival point of the crossing. The main points of interest include the Goat Hill Overlook, which Washington used as an observatory, and the Johnson Ferry House, which Washington likely visited on the way to Trenton.

---

34. INDEPENDENCE NATIONAL HISTORICAL PARK

143 South 3rd Street, Philadelphia, Pennsylvania, 19106

Site Type: Monuments – Independence Hall, Liberty Bell, Tomb of the Unknown Revolutionary War Soldier
Conflict: American Revolution
Dates: Hall completed in 1753
Designations: National Historic Park, National Historic District, UNESCO World Heritage Site
Web: www.nps.gov/inde (official website)

Independence National Historical Park, which incorporates over two dozen buildings in the heart of Philadelphia, is not strictly speaking a site of military interest. However, some of these buildings, most notably Independence Hall, are closely associated with the events of the American Revolution. In addition to Independence Hall, where the Declaration of Independence was signed, the park includes the Tomb of the Unknown Revolutionary War Soldier at nearby Washington Square.

In the years leading up to the American Revolution the city of Philadelphia was the site of the meetings of the Continental Congress, making it the de-facto capital of the rebels. It was here in 1776 that representatives of the thirteen American Colonies gathered to declare independence from, and by extension war on, Great Britain.

Although occupied by the British during part of the Revolution, Philadelphia was the heart of the colonies during and immediately after the war and served as the capital of the new nation until 1800. Independence National Historic Park was named a National Landmark District in 1966 and a UNESCO World Heritage Site in 1979.

The main site of interest is Independence Hall. Once the home of the Pennsylvania State Legislature, it has largely been preserved the way it looked in the late 18th century and is now maintained as a museum. The Liberty Bell, which once hung in the hall, is now on display in its own museum just outside. A few blocks south is Washington Square, a former city burial ground, where the Tomb of the Unknown Revolutionary War Soldier can be found. This monument honors all of those who fought in the War of Independence.

---

35. MUSEUM OF THE AMERICAN REVOLUTION

101 South 3rd Street, Philadelphia, Pennsylvania, 19106

Site Type: Museum
Conflicts: American Revolution
Dates: Opened in 2017
Web: www.amrevmuseum.org (official website)

The Museum of the American Revolution is the primary institution in the United States dedicated exclusively to the entirety of the Revolutionary War. Exhibits trace the history of the conflict, from unrest fomented during the Seven Years War through the final defeat of the British. While not part of Independence National Historical Park, the museum is just a few blocks east of Independence Hall and makes a good addition to a tour of the area.

The American Revolution took place from 1775 to 1783, with nearly a decade of unrest preceding the conflict. Events such as the Boston Massacre and Boston Tea Party led to a full-blown outbreak of hostilities in Massachusetts in the 1775 and the Declaration of Independence in 1776. Six years of war, first in the northern colonies and then in the southern colonies, culminated in defeat of the main British army at Yorktown in 1781. The war formally ended with the Treaty of Paris and American independence in 1783.

The Museum of the American Revolution is a huge facility with a number of exhibit halls which cover various aspects and time periods of the war. There are thousands of artifacts from the war on display, including items associated with George Washington and other military leaders of the rebellion. There is even a replica of the famous Boston Liberty Tree replete with lanterns. A highlight is the exhibit of extremely rare photos of American veterans of the Revolution taken in their later years.

---

36. USS OLYMPIA & INDEPENDENCE SEAPORT MUSEUM

211 South Columbus Boulevard, Philadelphia, Pennsylvania, 19106

Site Type: Cruiser
Conflicts: Spanish-American War, World War I
Dates: Commissioned in 1895
Web: www.phillyseaport.org (official website)

The Independence Seaport Museum is home to the USS Olympia, one of the only surviving American capital naval ships dating from the 19th century. In addition to the Olympia, the museum is also home to the submarine USS Becuna and a variety of exhibits on the maritime history of Philadelphia as well as life along the Delaware River.

The USS Olympia is one of the oldest surviving United States naval vessels, and possibly the only example of a protected cruiser still in existence, a protected cruiser being only partially armored in vital areas. Commissioned in 1895, the Olympia was stationed in the Pacific when war with Spain broke out. She was the flagship of the American force that destroyed the Spanish fleet at the Battle of Manila Bay in 1898.

The Olympia was decommissioned and recommissioned twice before seeing service in World War I and then the Russian Civil War. She carried the remains of the Unknown Soldier to the United States in 1921 before being permanently decommissioned the next year. The

Olympia was converted into a museum ship in the 1950s and became part of the Independence Seaport Museum in the 1990s.

The Independence Seaport Museum opened in 1961 and later became home to the USS Olympia and the submarine Becuna. Also to be found here is the Diligence, a recreation of a wooden schooner, which was used by the American Navy during the late 18th century. Another important exhibit of naval interest is Patriots and Pirates, which documents the history of America's fight against piracy in its early history.

---

37. VALLEY FORGE NATIONAL HISTORICAL PARK

1400 North Outer Line Drive, King of Prussia, Pennsylvania, 19406

Site Type: Encampment
Conflict: American Revolution
Dates: Camp established in 1777
Designations: National Register of Historic Places, National Historic Park, National Historic Landmark District, Pennsylvania State Park
Web: www.nps.gov/vafo (official website)

Valley Forge was the site of the famous winter encampment of George Washington and the Continental Army from late 1777 through the spring of 1778. The winter at Valley Forge was a critical event in the development of the American army and one that looms large in the imagination of the American people. Valley Forge, more than any other place, represents the fortitude and determination of those who fought in the War of Independence.

Following the reasonably successful campaigning season of 1777, in which the Americans enjoyed a major victory at Saratoga, the Continental Army was forced to retire after failing to drive the British from Philadelphia. They set up camp at Valley Forge where they spent the next few months subject to cold, food shortages and disease. By the end of the winter an estimated two thousand men, approximately one sixth of the army, had died, and most of the rest were unfit for service.

By spring, improved conditions, aid from France and the arrival of Baron von Steuben, a Prussian who came to assist in training the American troops, turned things around, and in June the rejuvenated army occupied Philadelphia. Valley Forge was named a Pennsylvania State park in 1893, incorporated into the Valley Forge Historic Landmark District in 1961, listed on the National Register of Historic Places in 1966 and named a National Historic Park in 1976.

Valley Forge National Historic Park preserves most of the site of the Continental Army's encampment. Highlights of the park include the National Memorial Arch, which commemorates those who served here in the Continental Army, as well as several reconstructed and recreated buildings. Among these is a single replica of the thousands of cabins that once dotted the encampment. There is a visitor's center with exhibits on the Valley Forge camp and the American Revolution.

---

38. AMERICAN MILITARY EDGED WEAPONRY MUSEUM

3562 Old Philadelphia Pike, Intercourse, Pennsylvania, 17534

Site Type: Museum
Dates: Opened in 1985
Web: www.discoverlancaster.com/directory/american-military-edged-weaponry-museum (official county tourism website)

The American Military Edged Weaponry Museum is possibly the most unexpected museum possible to be found in peaceful Amish Country. By contrast the town's other major museum is on quilting. However this stately and historic red brick building houses one of the finest collections of military knives and other bladed weapons in the United States.

Begun as a private collection, the museum features weaponry from most American conflicts spanning a century, beginning with the

Spanish-American War and continuing through the World Wars, the Cold War and into the present day. Although the exhibiting area is not huge, every display case and table is packed with a selection of knives, bayonets and swords and everything in between.

The American Military Edged Weaponry Museum's collection actually does go beyond edged weapons. There are also medieval arms on display, including such classic weapons as pikes and crossbow equipment. There is also a selection of firearms, and higher up on the walls, rounding out the exhibit, is an impressive collection of military-related artwork including original wartime propaganda posters.

---

39. NATIONAL CIVIL WAR MUSEUM

1 Lincoln Circle, Harrisburg, Pennsylvania, 17103

Site Type: Museum
Conflict: American Civil War
Dates: Opened in 2001
Web: www.nationalcivilwarmuseum.org (official website)

The National Civil War Museum is one of the largest museums related to the American Civil War in the United States. Founded by former Harrisburg mayor Steve Reed from a personal collection of Civil War artifacts, the museum focuses almost entirely on the military aspects of the conflict without political bias. The National Civil War Museum also doubles as the headquarters of the Sons of Union Veterans of the Civil War.

The museum is arranged in a series of galleries that trace the history of the conflict interspersed with exhibits ranging from weaponry to wartime medicine to music to Abraham Lincoln. There are over twenty thousand pieces in the collection, including personal items owned or used by Union leaders George McClellan, Winfield Scott and Ulysses S. Grant; and Confederate leaders J.E.B. Stuart, George Pickett, Stonewall Jackson and Robert E. Lee.

The National Civil War Museum's collection also includes artifacts from other related events, including John Brown's raid on Harper's Ferry in 1859, one of the major events that directly lead to the outbreak of the war; and Abraham Lincoln's assassination in 1865, the pivotal event of the war's aftermath. The museum also houses a large collection of manuscripts and letters related to the conflict.

---

40. GETTYSBURG NATIONAL MILITARY PARK

1195 Baltimore Pike, Gettysburg, Pennsylvania, 17325

Site Type: Battlefield - Battle of Gettysburg
Conflict: American Civil War
Dates: Battle fought on July 1-3, 1863
Designations: National Register of Historic Places, National Military Park
Web: www.nps.gov/gett (official website)

The Battle of Gettysburg was the largest battle ever fought on American soil, and one of the largest to take place anywhere during the 19th century. Fought in July of 1863, Gettysburg was the farthest that a southern army ever advanced into northern territory, and the field is often referred to as the Confederate high water mark. It was arguably the pivotal engagement of the war, forcing the South onto the defensive for the remainder of the conflict. A few months later Abraham Lincoln delivered his famous Gettysburg Address at the battle site.

In the summer of 1863 Robert E. Lee was presented with an opportunity to invade the northern states following his victory at Chancellorsville. Making his way up through Maryland into Pennsylvania, his strategic objective was to threaten and possibly cut off the key cities of Baltimore, Philadelphia and Washington DC. An even larger Union army under George Meade moved to stop Lee's march north, and the two sides clashed outside of the small Pennsylvania town of Gettysburg.

During the initial maneuvers the Union army beat the Confederates to the high ground. After two days of violent clashes, the decisive engagement came on July 3 in what is known as Pickett's Charge. This famous charge failed to dislodge the Union army from the heights and resulted in massive casualties for the Confederacy. The next day Lee was forced to withdraw back to Virginia, ending the Confederate invasion of the North. The Gettysburg Battlefield was designated a National Military Park in 1895 and listed on the National Register of Historic Places in 1966.

Gettysburg National Military Park is one of the most visited battlefields in the United States. It is especially revered among history buffs and military enthusiasts. The battlefield, just south of the city, is very well marked, and scores of monuments commemorate the soldiers of both sides. The wide open field where Pickett's Charge took place has been well preserved, with fences and other landmarks restored. Other popular sites include Big Round Top and Little Round Top; the Devil's Den; Cemetery Ridge and Seminary Ridge. The park also incorporates the Gettysburg National Cemetery.

---

41. FORT NECESSITY NATIONAL BATTLEFIELD

1 Washington Parkway, Farmington, Pennsylvania, 15437

Site Type: Battlefield - Battle of Fort Necessity
Conflict: French and Indian War
Dates: Battle fought on July 3, 1754
Designations: National Register of Historic Places, National Battlefield
Web: www.nps.gov/fone (official website)

The Battle of Fort Necessity is one of those events that many people have heard of but know little about. An engagement of the French and Indian War, it is perhaps best known for its association with the early military career of George Washington. The battle was named for a wooden stockade that was erected in haste as a defense for the British in the face of a much larger French force.

During the early stages of the French and Indian War there were clashes between the British and French on the frontier of the American colonies. These played little strategic role in the greater conflict, as most of the action of the war took place in Upstate New York and Quebec. However, it did foreshadow the future conflict between the American colonists and the Native Americans of the Ohio River Valley.

The most famous of these battles took place in Western Pennsylvania in July of 1754. George Washington, a colonel at the time, led a force of about four hundred men to intimidate the French into leaving the area. The plan backfired, and a superior French force attacked the British. Despite the erection of a makeshift stockade, the British were defeated and forced to leave. However the battle made George Washington a household name and became part of his legend. Fort Necessity was designated a National Battlefield in 1931 and listed on the National Register of Historic Places in 1966.

Fort Necessity National Battlefield is an extensive open ground incorporating most of the area in which the battle took place. A replica of the fort has been erected on the site of the original stockade. Also located at Fort Necessity is the gravesite of Edward Braddock, a British general who died during the campaign to take Fort Duchesne in 1755.

---

42. WOODVILLE PLANTATION

1375 Washington Pike, Bridgeville, Pennsylvania, 15017

Site Type: Battlefield – Battle of Bower Hill
Conflict: Whiskey Rebellion
Dates: Skirmish fought on July 16, 1791
Web: www.woodvilleplantation.org (official website)

Woodville Plantation was the site of a skirmish of one of the earliest conflicts in American history: the Whiskey Rebellion. This extremely minor event, which occurred in Western Pennsylvania in 1791, looms large in the minds of many Americans who fear the overreach of the federal government in the lives of its citizens. Woodville Plantation, where the rebellion nominally began, was the site of the only fighting of the uprising.

In the years immediately following the ratification of the United States Constitution, one of the first duties of the new government was to develop sources of revenue. One of these was a tax on whiskey, one of the most popular commodities at the time. The producers of whiskey in Pennsylvania, many of whom had just fought a war against unfair taxation, generally refused to pay, thereby launching the Whiskey Rebellion.

In actuality a series of demonstrations, protests and a general refusal to pay the tax, there was little real fighting. The only exception was the minor Battle of Bower Hill, fought on July 16, 1791. At least two and perhaps four men were killed, and a federal marshal who was attempting to collect the tax was driven from the area. Eventually a strong force of soldiers and militiamen marched on Western Pennsylvania and dispersed the rebels.

The Woodville Plantation where some of the events of the Whisky Rebellion took place is still in existence. At its heart is the Neville House. Built in 1785, the house is preserved much as it was in the 18th century. An historical marker on the site of the plantation commemorates the events of the tiny but memorable uprising.

MID-ATLANTIC

43. FORT DELAWARE STATE PARK

45 Clinton Street, Delaware City, Delaware, 19706

Site Type: Masonry Fort
Conflict: American Civil War
Dates: Originally completed in 1868
Designations: National Register of Historic Places, Delaware State Park
Web: www.destateparks.com/history/fortdelaware (official website)

Fort Delaware was one of the last major coastal fortifications constructed by the United States in the years leading up to the Civil War. It was one of America's longest-used forts and was still active during the World Wars. Never directly attacked, Fort Delaware was used primarily as a military prison during the Civil War. It is now the main point of interest of Fort Delaware State Park and Pea Patch Island.

The strategic value of Pea Patch Island in the Delaware River was recognized as early as the 18th century. An early fort was constructed on the island in the 1820s, but this was badly damaged by fire in 1831 and torn down a few years later. Construction on the current fort began in 1848 and continued on and off into the 1860s. Fort Delaware was operational at the outset of the American Civil War.

Once it became clear that the Confederate navy posed little threat to the area, Fort Delaware was converted to use as a military prison. Thousands of Confederate soldiers were incarcerated here during the war, including many that surrendered at the Battle of Gettysburg. The fort remained garrisoned during the Spanish-American War, World War I and World War II. The fort was closed in the postwar era and turned over to the state of Delaware. It was listed on the National Register of Historic Places in 1971.

Fort Delaware is largely intact today, despite being damaged by several storms in the late 19th century. A masonry structure with a water-filled moat that is still in use, the fort is accessible only by a single bridge. Inside are various barracks that were added during the Civil War. There are two nicer red-brick buildings used to house guards and prisoners who were officers; and a dark and dreary general barracks used to house privates and non-commissioned officers. The fort is now run as a living museum.

---

44. FORT MCHENRY NATIONAL MONUMENT

2400 East Fort Avenue, Baltimore, Maryland, 21230

Site Type: Masonry Star Fort
Conflict: War of 1812
Dates: Originally completed in 1800
Designations: National Monument
Web: www.nps.gov/fomc (official website)

Fort McHenry was one of the earliest military fortifications constructed by the United States government. Built to protect the city of Baltimore from naval attack, the fort was put to the test a little more than a decade after its completion during the War of 1812. It was during a British naval attack against the fort that Francis Scott Key wrote the poem that was later immortalized as the American National Anthem, the *Star Spangled Banner*.

Baltimore Harbor was protected by fortifications as early as the 1770s. In the late 1790s, the old Fort Whetstone was demolished and replaced with a much larger and stronger structure, Fort McHenry. Designed to withstand heavier artillery fire from more modern naval guns, the new fort played a pivotal role in the War of 1812. In September 1814 a British fleet bound for Baltimore was forced to stop and deal with Fort McHenry before proceeding on to the city. This was less than a month after the British sacked and burned Washington DC, so the fort's defenders took the threat very seriously.

On September 13-14, 1814, the Americans endured a withering bombardment from the British navy. Amazingly the fort suffered only light damage and few casualties, and the invasion fleet was forced to turn back. This battle, one of the few major American victories of the War of 1812, was immortalized by Francis Scott Key, who witnessed the engagement from aboard a nearby ship. The fort subsequently served as a prison during the Civil War, as an army hospital during World War I, and as a coast guard base during World War II. Fort McHenry was designated a National Monument in 1925.

Fort McHenry was constructed towards the end of the Napoleanic era, when artillery was becoming too powerful for masonry fortifications to withstand. Although built along the classic star-fort pattern, it relies much more heavily on earthworks designed to absorb gunfire. Because it was used for a variety of purposes up until the middle of the 20th century, Fort McHenry has been fairly well preserved. The fort's museum boasts a collection of famous American flags. Unfortunately the one that inspired the national anthem is not here, but at the Smithsonian Institute in Washington, DC.

---

45. USS CONSTELLATION

301 East Pratt Street, Baltimore, Maryland, 21202

Site Type: Wooden Sloop
Conflict: American Civil War
Dates: Commissioned in 1855
Designations: National Register of Historic Places, National Historic Landmark
Web: www.historicships.org (official website)

The USS Constellation was one of the largest wooden warships ever built by the United States Navy. Completed in 1854, it saw much use in the years leading up to the American Civil War, primarily in disrupting the slave trade. From the end of the Civil War until its decommissioning in 1933 it served as a training ship, a supply ship, and as a diplomatic ship. It later enjoyed a brief revival as honorary flagship in the Atlantic theater during World War II. It was decommissioned and re-designated as a museum ship in the 1950s.

The Constellation was one of the most powerful wooden-hull warships ever constructed. Entering service in 1855, the Constellation spent the years leading up to and during the American Civil War largely patrolling the Atlantic. She was active in disrupting the slave trade, seizing several ships and freeing many soon-to-be slaves in the years prior to the war. After the South seceded, the Constitution spent much of the war across the Atlantic chasing down Confederate commerce raiders.

After the Civil War the Constellation remained in active service for nearly a century, primarily as a training vessel, especially during World War I. She also ran emergency food supplies to Ireland in the 1870s. During World War II the Constellation served as the honorary reserve flagship of America's Atlantic fleet. The USS Constellation was designated a National Historic Landmark in 1963 and added the National Register of Historic Places in 1966.

The USS Constellation received its final decommission in 1955, a little more than a century after its completion. A few years later it was permanently moored in Baltimore harbor as a museum ship. It was restored again in the 1990s, and is now the only surviving American naval warship from the mid-19th century. It is part of the Historic Ships of Baltimore Museum, which is also home to the submarine USS Torsk and other ships.

---

46. UNITED STATES NAVAL ACADEMY

121 Blake Road, Annapolis, Maryland, 21402

Site Type: Military Academy
Dates: Founded in 1845
Web: www.usna.edu (official website)

The United States Naval Academy, commonly referred to as Annapolis, is the primary college for the training of American Navy officers. Established in 1845, it is the second oldest of America's professional military academies. Although most of the campus is not open to visitors, especially when school is in session, there is a visitor center, and limited tours of the academy are available.

The Naval Academy stands on a site formerly used as an army base. The school was founded on a limited basis with only a handful of students. The academy was reorganized and expanded in the 1850s, just in time to prepare naval officers for the American Civil War. Graduates of the academy served in both the Union and Confederate navies.

The importance of the navy grew considerably during and after the Spanish-American War, and coursework at the academy expanded significantly. By the time World War I broke out, graduates of the academy made up much of the naval officer corps. Notable alumni include Chester Nimitz, Ross Perot, John McCain, Alan Shepard and Jimmy Carter.

The United States Naval Academy can be visited, with tours beginning at the Armel-Leftwich Visitor's Center, which has exhibits on the academy and cadet life. The Naval Academy Museum is home to, among other things, the Rogers Collection of model ships. There are also numerous monuments and memorials located throughout the campus, including the Tripoli Monument, the oldest military monument in the United States, commemorating America's victory over the Barbary Pirates.

---

47. PATUXENT RIVER NAVAL AIR MUSEUM

22156 Three Notch Road, Lexington Park, Maryland, 20653

Site Type: Museum
Dates: Opened in 1975
Web: www.paxmuseum.com (official website)

The Patuxent River Naval Air Museum is an aviation museum associated with the Patuxent River Naval Air Station. After the National Naval Aviation Museum in Pensacola it is one of the largest such museums in the United States. It houses collections from several naval air bases across the country, some of which are no longer in operation.

The Naval Air Station at Patuxent River traces its roots back to an early testing ground established in 1937. A full-fledged base was established in 1942 following America's entry into World War II. Throughout the Cold War it was used as a training facility for test pilots and aircraft development. Several other naval air stations in the Mid-Atlantic area were consolidated here in the 1990s.

The Patuxent River Naval Air Museum was opened in 1975 to showcase the history of the base as well as aircraft and other artifacts. Additional exhibits were brought here from other bases after the end of the Cold War. The collection includes aircraft mostly from the postwar era. The museum is also home to flight simulators that re-create the experience of flying in an F-14 aircraft.

---

48. MONOCACY NATIONAL BATTLEFIELD

5201 Urbana Pike, Frederick, Maryland, 21704

Site Type: Battlefield – Battle of Monocacy
Conflict: American Civil War
Dates: Battle fought on July 9, 1864
Designations: National Register of Historic Places, National Historic Landmark, National Battlefield
Web: www.nps.gov/mono (official website)

The Battle of Monocacy was the last engagement of the last Confederate attempt to invade the northern states and capture Washington DC. While a victory for the South, which brought the Confederacy within sight of the Union capital, the rebels had been both delayed and reduced in number. This made the capture of Washington virtually impossible, and the rebels were forced to withdraw to Virginia.

In the early summer of 1864 a Confederate force under the command of Jubal Early managed to capture several important cities in Virginia and Maryland that offered a route into Washington, DC, all while the Union was preparing a major foray into Virginia. Concerned at this unexpected development, the North sent a force under Lew Wallace to halt the rebel advance. Underestimating the strength of the Confederacy, the Union army was significantly outnumbered. Despite this, Wallace met the invaders at Monocacy Junction.

The fighting was ferocious, with significant casualties on both sides, and in the end the Union army was forced to retreat. However, the engagement cost the Confederacy a day, enough time to strengthen the Union defenses. Faced with poor prospects of victory, Early was forced to march his army back to Virginia. Monocacy was listed on the National Register of Historic Places and named a National Historic Landmark in 1973, and designated a National Battlefield in 2001.

Monocacy National Battlefield is probably one of the less well known Civil War battle sites, due in part to the fact that little attention was paid to the area until late in the 20th century. Only part of the battlefield is within the confines of the park, and many of the commemorative stone markers were privately commissioned and placed.

---

49. NATIONAL MUSEUM OF CIVIL WAR MEDICINE

48 East Patrick Street, Frederick, Maryland, 21701

Site Type: Museum
Dates: Opened in 1996
Web: www.civilwarmed.org (official website)

The National Museum of Civil War Medicine is devoted to telling the story of medical practices during America's greatest military conflict. The Civil War, considered by many historians to be the first true modern war, witnessed the beginning of many modern battlefield medical practices, at least as far as was available at the time. By the end of the war, the medical profession in the United States had changed forever.

The museum began as a private collection of medical instruments courtesy of Dr. Gordon Damman. Damman spearheaded an effort to turn his collection into a gallery that would be accessible to the public.

The National Museum of Civil War medicine was opened in 1996.

The National Museum of Civil War Medicine collection has grown significantly over the last two decades. A large number of artifacts are displayed in a series of exhibits that include everything from battlefield care to military hospitals of the time. The museum now operates several other facilities, including the Pry House Field Hospital Museum in Sharpsburg.

---

50. ANTIETAM NATIONAL BATTLEFIELD

5831 Dunker Church Road, Sharpsburg, Maryland, 21782

Site Type: Battlefield - Battle of Antietam
Conflict: American Civil War
Dates: Battle fought on September 17, 1862
Designations: National Register of Historic Places, National Battlefield
Web: www.nps.gov/anti (official website)

The Battle of Antietam, also known as the Battle of Sharpsburg, was one of the most important and bloodiest battles of the American Civil War. The culmination of Robert E. Lee's first major effort to invade the North in 1862, Antietam is considered to have been inconclusive. However, since Lee was forced to withdraw his army afterwards, it was technically a Union victory. Antietam is infamous due to the fact that more American soldiers were killed in a single day here than in any other battle in United States history.

In the summer of 1862, following their victory at the Second Battle of Bull Run, the Confederate Army of Northern Virginia set its sights on an invasion of the northern states. In September Lee marched his army into Maryland. A few days later, a larger Union army led by George McClellan marched out from Washington DC to meet them. In mid-month there were some initial clashes, with the cautious McClellan famously passing on opportunities to do heavy damage to Lee's divided army.

On September 17 the two armies met at Antietam. Although the Confederate army was significantly outnumbered they managed to fight the Union army to a draw. However, the resultant casualties were catastrophic on both sides. The North lost over twelve thousand men, while the South lost a little over ten thousand. Overall the losses were more devastating to the Confederacy, and Lee withdrew, thus ending his first major northern offensive. The Antietam Battlefield was declared a National Battlefield in 1890.

Antietam National Battlefield is spread out over a large area just west of Frederick. It includes the entirety of the area where the battle took place, including a sprawling cemetery where many of the Union dead are buried, and the Pry House, which was the headquarters of General McClellan. There is also a visitor's center with exhibits on the battle and Abraham Lincoln's battlefield visit.

---

51. DISTRICT OF COLUMBIA WAR MEMORIAL

West Potomac Park, Independence Avenue SW, Washington, DC, 20024

Site Type: Monument – World War I
Conflict: World War I
Dates: Dedicated in 1931
Designations: National Register of Historic Places
Web: www.nps.gov/articles/the-district-of-columbia-war-memorial.htm (official website)

World War I was the conflict that heralded America's entry into the modern age as a true world power. Nearly five million Americans served in uniform during the War to end all Wars, ultimately tipping the scale in favor of Britain and France against Germany. Over fifty thousand servicemen were killed during the war, the third highest of any conflict in American history.

In 1924 Congress commissioned a memorial for the veterans of World War I. It is one of two monuments in Washington DC commemorating those who served in the Great War, the other being the Navy-Merchant Marine Memorial. It was completed in 1931 and dedicated by President Herbert Hoover on Veteran's Day. The memorial was renovated and expanded in 2011 and listed on the National Register of Historic Places in 2014.

The District of Columbia War Memorial is an open-air domed rotunda in West Potomac Park just off of the National Mall. According to records, four hundred and ninety nine servicemen from the Washington DC area were killed in the conflict. Their names are inscribed around the monument. The names of all of the local area veterans of World War I are preserved in the cornerstone.

---

52. NAVY-MERCHANT MARINE MEMORIAL

Lady Bird Johnson Park, George Washington Memorial Parkway, Washington, DC, 20037

Site Type: Monument – United States Navy & Merchant Marines
Conflict: World War I
Dates: Dedicated in 1934
Web: www.nps.gov/gwmp/planyourvisit/ladybirdjohnsonpark.htm (official website)

World War I saw the debut of America's modern navy on the world stage. First begun in the late 19th century, and tried out during the Spanish-American War, the American Navy was becoming competitive with the greatest of the European powers when World War I broke out. Although military action in the Atlantic theater was minimal during the Great War, both the navy and merchant marines were honored for their service, especially against German U-Boats.

The Navy-Merchant Marine Memorial was begun shortly after World War I ended, and it was completed and dedicated in 1934. Unlike most of the other national military memorials which are located on or near the National Mall, the Navy-Merchant Marine Memorial is located across the Potomac River in Lady Bird Park on Columbia Island.

The Navy-Merchant Marine Memorial is a really unique feature of the Washington DC landscape. Unlike most of the other war monuments, which typically feature large granite structures and/or bronze statues of servicemen and other heroes, this monument features an immense sculpture of a cresting wave crowned by a flock of seven seagulls. The commemoration is somewhat minimalist and general, recalling all of those who served at sea during the War to end all Wars.

---

53. NATIONAL WORLD WAR II MEMORIAL

1750 Independence Avenue SW, Washington, DC, 20024

Site Type: Monument – World War II
Conflict: World War II
Dates: Dedicated in 2004
Web: www.nps.gov/wwii (official website)

World War II was the most immense and terrible war in the history of mankind. Spanning every corner of the globe and almost incomprehensible in its destructiveness, it has been estimated that over sixty million people lost their lives during the conflict. Over sixteen million Americans served in World War II, almost as many as all of its other wars put together, and more Americans died in this conflict than in any other war except the Civil War.

It took nearly half a century for the United States government to begin the process of commemorating the huge and unparalleled undertaking. Congress did not officially pass legislation for the establishment of the National World War II Memorial until 1993, and it was nearly another decade further before work actually began. This long awaited monument was finally completed and dedicated in 2004, nearly sixty years after the war's ending.

The National World War II Memorial is absolutely massive, dwarfing all of the other military monuments located on the National Mall. The footprint, located not too far from the Washington Monument, is immense. There are many elements to the monument, including two vast wings representing the European and Pacific Theaters of the war; granite monoliths representing the fifty states; countless memorial plaques; and a wall of over four thousand gold stars, one for every one hundred American servicemen who died during the conflict.

---

54. KOREAN WAR VETERAN'S MEMORIAL

900 Ohio Drive SW, Washington, DC, 20024

Site Type: Monument – Korean War
Conflict: Korean War
Dates: Dedicated in 1995
Designations: National Register of Historic Places
Web: www.nps.gov/kowa (official website)

The Korean War was the first major conflict fought by the United States in the post-World War II era. It was also one of the only major police actions fought under the auspices of the United Nations. Well over a million American servicemen participated in the conflict, and over thirty thousand were killed by the time hostilities came to a close in 1953.

It wasn't until 1986, more than three decades later, that Congress finally authorized a national memorial to those who had served in Korea. Started under President George Bush Sr., it took nearly ten years to complete, and was dedicated in 1995 by President Bill Clinton. The Korean War Veteran's Memorial was listed on the National Register of Historic Places in 1995.

The Korean War Veteran's Memorial incorporates several major elements, most famously the series of nineteen immense steel soldiers cast in marching positions as if on an endless patrol. All four branches of the military are represented among the statues. There is a huge wall full of images from the war, and another commemorating the twenty two countries from the United Nations that participated in the conflict.

55. VIETNAM VETERAN'S MEMORIAL & WOMEN'S MEMORIAL

5 Henry Bacon Drive NW, Washington, DC, 20245

Site Type: Monument – Vietnam War
Conflict: Vietnam War
Dates: Dedicated in 1982
Designations: National Register of Historic Places, National Monument
Web: www.nps.gov/vive (official website)

The Vietnam Veteran's Memorial is arguably the most poignant military memorial on the National Mall, and the second most visited after the National World War II Memorial. Established in the wake of one of the most tragic chapters in American history, in which veterans of the highly unpopular war came home to an unforgiving populace, the memorial represents a terrible scar on the collective American psyche. The memorial includes a monument specifically dedicated to the women who served.

In 1975 the last Americans in Vietnam were forced to abandon their final positions in the capital of Saigon. The following decade was a period of great soul searching about why we had been there and what it had accomplished. It was also a period of disregard for those who had served in the war. In 1979 congress authorized the establishment of what would become the most heart-wrenching memorial in Washington, D.C. The Vietnam Veteran's Memorial was dedicated and listed on the National Register of Historic Places in 1982.

The Vietnam Veteran's Memorial is unlike anything else on the National Mall. An immense black wall nearly the length of a football field that cuts into the earth, the names of nearly sixty thousand Americans who died in Vietnam are carved into its face. Two statues were added to the memorial in later years: The Three Soldiers in 1984, honoring all of those who fought on the ground; and the Vietnam Women's Memorial in 1993, honoring the many women who served there in uniform.

56. UNITED STATES NAVY MEMORIAL & NAVAL HERITAGE CENTER

701 Pennsylvania Avenue NW, Washington, DC, 20004

Site Type: Monument – United States Navy
Dates: Dedicated in 1987
Web: www.navymemorial.org (official website)

The United States Navy Memorial honors all of those who have ever served in the United States Navy, the Coast Guard, the Merchant Marine and the Marine Corps, from the American Revolution through the present day. Unlike the other major memorials in Washington DC honoring the United States military, the Navy Memorial is also home to a heritage center, with revolving exhibits and event facilities.

A memorial to those who have served in the United States Navy was envisioned as early as the 19th century, but no serious action was ever taken until the late 1970s. Authorized by congress and started under Jimmy Carter in 1980, the first elements of the memorial were completed in 1987. The memorial was formally dedicated on October 13, 1987, while the heritage center was opened in 1991.

The United States Navy Memorial consists of a large plaza and building complex overlooking Pennsylvania Avenue. Elements on the plaza include the statue of The Lone Sailor, a huge fountain and immense ship's masts in lieu of traditional flagpoles flying traditional semaphore banners in addition to the flag. The two buildings that constitute the Naval Heritage Center house revolving exhibits of naval interest as well as the service records for the men and women who have served the United States at sea.

57. NATIONAL MUSEUM OF THE UNITED STATES NAVY

736 Sicard Street SE, Washington, DC, 20374

Site Type: Museum
Date: Opened in 1961
Web: www.history.navy.mil/content/history/museums/nmusn.html (official website)

The National Museum of the United States Navy is one of the oldest military museums in the United States, with a collection that can be traced back to the 19th century. Founded in 1961, the museum was preceded by earlier institutions dating back to the American Civil War. The museum and its growing collection was moved to a permanent home in the Washington Navy Yard in the 1960s and now serves as the flagship of American naval museums around the country.

The collection of the museum began with a French naval cannon captured around the year 1800. A formal effort to store and display naval artifacts began at the end of the Civil War, when space was set aside at the Washington Naval Yard in the capital. It was moved several times due to space needs, ultimately ending up at the old naval gun factory.

The National Museum of the United States Navy is home to an immense collection of naval artifacts. There are exhibits from the American Revolution, the Civil War, both World Wars and every military action in between. At the time of this writing there were special exhibits on the sinking of the USS Maine, the Byrd expedition to the South Pole and the history of America's submarine force.

58. SMITHSONIAN NATIONAL AIR AND SPACE MUSEUM

600 Independence Avenue SW, Washington, DC, 20560

Site Type: Museum
Dates: Opened in 1976
Web: https://airandspace.si.edu (official website)

The National Air and Space Museum at the Smithsonian Institute is the definitive aerospace museum of the United States, if not the world. The successor to an earlier National Air Museum, the Air and Space Museum was built both to accommodate the growing interest in space exploration as well as the growing collection of aircraft and related artifacts that were overflowing the original museum. A new building was constructed in the 1970s and filled with some of America's most historic air and spacecraft.

Although not a military museum per se, the Air and Space Museum has in its collection a number of famous and historic military aircraft. There are planes here from all of the service branches as well as a few from foreign countries. There are German Fokker and British Sopwith biplanes from World War I, Japanese Zero and American P-51 fighters from World War II, and many others. The Enola Gay, the bomber which dropped the atomic bomb on Hiroshima in 1945, was formerly on display but removed, in part due to the controversial nature of its famous mission.

In addition to the many military craft kept here, the museum houses some of the most famous aircraft and spacecraft in history, including the original 1903 Wright Brothers Flyer; the 1927 Spirit of St. Louis, the first plane to fly across the Atlantic; and the Apollo 11 command module, the space craft which took American astronauts to the first moon landing in 1969. An annex to the museum located near Dulles Airport houses well over a hundred additional aircraft.

59. SMITHSONIAN NATIONAL MUSEUM OF AMERICAN HISTORY

1300 Constitution Avenue NW, Washington, DC, 20560

Site Type: Museum
Dates: Opened in 1964
Web: https://americanhistory.si.edu (official website)

The National Museum of American History is not a military museum. However, documenting as it does the history of the United States, there are a number of exhibits of military interest that merit its inclusion as a must-see military site. One of the older museums of the Smithsonian Institute, it has been open since 1964 and is among the most visited museums located on the National Mall.

A visitor could spend days perusing the countless artifacts of Americana on display in this museum. But for those interested in military history, there are two focal points. The first is one of the museum's signature attractions: the original flag that flew over Fort McHenry during the War of 1812 and which is now known as *the* Star Spangled Banner. It was this flag that inspired Francis Scott Key to write the poem that would later go on to become the national anthem.

The Museum of American History's primary area of military interest is the east wing of the third floor, which is entirely devoted to military history. This wing features, among other things, the recovered USS Philadelphia, a gunboat that served and was sunk during the American Revolution. There are literally thousands of artifacts on display here, including examples of every conceivable small arm weapon that was ever used by the United States military, flags, uniforms, equipment and many other artifacts.

60. NATIONAL MUSEUM OF AMERICAN JEWISH MILITARY HISTORY

1811 R Street NW, Washington, DC, 20009

Site Type: Museum
Dates: Opened in 1958
Web: https://nmajmh.org (official website)

The National Museum of American Jewish Military History is one of America's most unique military commemorative institutions. Established by the Jewish War Veterans of the USA in the 1950s, it is located in that organization's national headquarters in Washington, DC. The museum was created to honor and tell the story of the Jewish veterans of America's many wars. It is part of the Dupont-Kalorama Museums Consortium.

Jews have served in uniform in almost every major conflict in American history. Over a hundred Jews fought for independence in the American Revolution, while perhaps as many as ten thousand fought in the Civil War, mostly for the Union, but a few for the Confederacy. Over half a million Jews fought for the United States during World War II, more than ten percent of the entire American Jewish population at the time. The Jewish War Veterans of the United States of America has been active since the late 19th century.

The National Museum of American Jewish Military Heritage was founded in 1958 and moved to its present location in 1984. The main exhibit tells the story of Jews in the American military from colonial times to the 21st century. There is a focus on World War II, due in part to the huge contribution of Jews fighting in that war as well as the particularly Anti-Semitic nature of the enemy. Also on site is an exhibit honoring Jewish veterans whose service was so distinguished they received the Congressional Medal of Honor.

61. CONGRESSIONAL CEMETERY

1801 E Street SE, Washington, DC, 20003

Site Type: Cemetery
Dates: Opened in 1807
Designations: National Register of Historic Places, National Historic Landmark
Web: www.congressionalcemetery.org (official website)

The Congressional Cemetery is the oldest national cemetery in the United States, and the only one that predates the Civil War. There are a considerable number of prominent and famous burials within the cemetery, including several of military importance. In addition to the many burials of those who have served in the armed forces, there are a number of important graves of civilians of military related interest. The Congressional Cemetery was listed on the National Register of Historic Places in 1969 and named a National Historic Landmark in 2011.

Probably the most famous military figure buried here is Joseph Gilbert Totten, who served in the army for much of the first half of the 19th century, and was for a time its chief engineer. Also of particular interest here is Archibald Henderson, who served as Commandant of the Marine Corps for nearly forty years. Arguably the two most interesting civilian figures buried in the Congressional Cemetery, from a military history perspective, are Matthew Brady and John Philip Sousa.

Matthew Brady, the most famous photographer of the 19th century, became renowned for the thousands of pictures he took that documented the American Civil War. He is known today as the world's first wartime photographer. John Philip Sousa was a longtime member and later conductor of the Marine Corps Band. One of the greatest composers of military marches in history, his work *Stars and Stripes Forever* is now the national march of the United States.

62. ANACOSTIA PARK

1900 Anacostia Drive SE, Washington, DC, 20020

Site Type: Event – Bonus March Riot
Dates: Riot took place on July 28, 1932
Web: www.nps.gov/anac (official website)

Anacostia Park was the site in of one of the most shameful moments in American military history. It was here in 1932, at the height of the Great Depression, that the United States Army took heavy handed action against veterans that had come to Washington to demand their bonus money. The action, which was organized by future World War II leaders MacArthur, Patton and Eisenhower, was almost universally condemned and contributed to Franklin Roosevelt's victory over Herbert Hoover in the presidential election later that year.

In 1924 Congress enacted a special bonus to be paid to veterans who had served in World War I, with a payout date of 1945. By the early 1930s many of these veterans were reeling from the ravages of the Great Depression, and began to inquire into an earlier payout of the bonus. In June of 1932 thousands of veterans, many with family members in tow, descended on Washington to demand the payout of the bonus. After Congress failed to authorize this, the veterans, who came to be known as Bonus Marchers, established a Hooverville camp in Anacostia Park.

On July 28 President Hoover ordered the camp dispersed. A police shooting led to the deaths of two veterans, and rioting ensued. The army was called in, and by the end of the day the veterans were dispersed. However over a thousand veterans, policemen and soldiers had sustained injuries. The public response was furious, and a few months later Hoover lost his re-election. Congress authorized the early payment of the bonus in 1936.

Anacostia Park is maintained as a public park by the National Park Service. Strangely there is little in the way of public information on the Bonus Army here, and very little indication of anything commemorating the event. As of this time of this writing there were no known plans for a monument or any other memorial at Anacostia Park.

---

63. THE PENTAGON

1400 Defense Boulevard, Washington, DC, 20301

Site Type: Military Command Center
Dates: Opened in 1943
Designations: National Register of Historic Places, National Historic Landmark
Web: https://pentagontours.osd.mil/Tours (official website)

The Pentagon is the commonly used name for the building which houses the United States Department of Defense. While technically located in Arlington, Virginia, it has a Washington, DC address. An absolutely sprawling office complex with five sides, which inspired its nickname, it is the location of the office of the Secretary of Defense as well as civilian staff, enlisted men and officers of all branches of the United States Military.

For most of its history the Department of Defense and its predecessors was housed in a series of small and highly inadequate buildings in Washington, DC proper. With World War II looming it was decided that a much bigger and more permanent facility was needed. Following a short stay in the building that now houses the State Department, the Pentagon was formally opened in 1943.

With its completion the Pentagon became the nerve center for all American military operations, first during World War II, and later during the Cold War. It famously made the news in 2001 as one of the terrorist targets of 9/11. Nearly two hundred people were killed in the attack including many Pentagon personnel. The Pentagon was listed on the National Register of Historic Places in 1989 and named a National Historic Landmark in 1992.

The Pentagon has offered tours since the 1970s with occasional periods of interruption, notably following the terrorist attacks of 9/11 and during the Coronavirus outbreak. As of the time of this writing visitors were allowed on tours that require reservations at least two weeks in advance. The tour highlight is arguably the Hall of Heroes, which honors those who have served in the military and who have received the Congressional Medal of Honor.

VIRGINIA

64. ARLINGTON NATIONAL CEMETERY

1 Memorial Avenue, Arlington, Virginia, 22211

Site Type: Cemetery
Dates: Opened in 1864
Designations: National Register of Historic Places, National Cemetery
Web: www.arlingtoncemetery.mil (official website)

Arlington National Cemetery is the best known and most visited of America's many military cemeteries. Established on land that had formerly been the estate of Robert E. Lee, it is one of the largest cemeteries in the United States and is home to many military monuments. Entire sections have been established for specific wars and military service, with special areas designated for nurses, chaplains and many other groups, making Arlington one of the world's most comprehensively organized military burial grounds.

The estate where Arlington Cemetery now stands belonged to the descendents of Martha Washington in the early 19th century and was later passed down to Robert E. Lee. The land was seized by the federal government during the Civil War, though Lee's descendents later received compensation. Burials at the site began in 1864 to accommodate the thousands of soldiers who died at the Battle of the Wilderness. The cemetery has been expanded on several occasions over the last century and a half.

The sprawling cemetery is home to over four hundred thousand graves and counting. Two United States presidents, William Howard Taft and John F. Kennedy, are buried at Arlington, the latter marked by the cemetery's famous eternal flame. Many other political and military leaders are buried here, as well as many American heroes such as astronauts Gus Grissom and John Glenn. The most honored site is the Tomb of the Unknown Soldier. This tomb, perpetually guarded, contains the remains of four unidentified solders who served in World War I, World War II, Korea and Vietnam respectively.

Arlington National Cemetery is packed with hundreds of memorials and monuments honoring everything from famous military units such as the Buffalo Soldiers and Rough Riders to special units like military police and even civilian journalists. There are memorials to those who have died in terrorist attacks, space shuttle disasters, and many battles. The greatest monument here, and one of the most photographed statues in the world, is the Marine Corps War Memorial, featuring six marines raising the American flag at the top of Mount Suribachi during the Battle of Iwo Jima in World War II.

---

65. UNITED STATES AIR FORCE MEMORIAL

1 Air Force Memorial Drive, Arlington, Virginia, 22204

Site Type: Monument – United States Air Force
Dates: Dedicated in 2006
Web: www.afdw.af.mil/afmemorial (official website)

The United States Air Force Memorial honors all of those who have ever served in the United States Air Force, its predecessor the Army Air Corps, and other related military departments. One of the most recent major military monuments to be erected in the Washington DC area, it is also one of the few not located on the National Mall.

A foundation to support a national Air Force memorial was formed in the early 1990s, just a few years after the National Navy Memorial was dedicated in the District of Columbia. Authorized by Congress and started under George W. Bush in 2001, the memorial took just a few years to complete and was dedicated on October 14, 2006.

The United States Air Force Memorial is located close to both Arlington National Cemetery and the Pentagon. Its primary feature is a trio of huge futuristic towers designed to look like the flying bomb burst formation. Walls at the base of the spires are inscribed with the names of those who served in the Air Force and who received the Congressional Medal of Honor.

66. NATIONAL MUSEUM OF THE MARINE CORPS

18900 Jefferson Davis Highway, Triangle, Virginia, 22172

Site Type: Museum
Dates: Opened in 2006
Web: www.usmcmuseum.com (official website)

The National Museum of the Marine Corps is one of the only museums in the United States specifically dedicated to the Marines. It is located near the Quantico Marine Corps Base, one of America's most famous military installations. Completed in 2006, the museum traces the history of the United States Marine Corps with a focus on their activities in the 20th century.

The Marines have been around in one form or another since the late 1700s. Playing a role in almost every major American war since the Revolution, they rose to prominence as an elite force during the Spanish-American War. In 2006 the National Museum of the Marine Corps opened, replacing two smaller earlier museums at Quantico and the Washington Navy yard.

The National Museum of the Marine Corps is home to over half a dozen major exhibits, with several others on the way. Some of the galleries focus on the life of Marines, while others trace their service record through America's wars. As of the time of this writing new exhibits were being planned related to the Marine Corps in the present day.

67. MANASSAS NATIONAL BATTLEFIELD PARK

6511 Sudley Road, Manassas, Virginia, 20109

Site Type: Battlefield – First and Second Battles of Bull Run
Conflict: American Civil War
Dates: Battles fought on July 21, 1861 & August 28-30, 1862
Designations: National Register of Historic Places, National Battlefield Park
Web: www.nps.gov/mana (official website)

Manassas National Battlefield Park was the site of two major engagements of the American Civil War. The first, which took place on July 21, 1861, was known as the First Battle of Bull Run, and was an unexpected victory for the South thanks to the heroics of Thomas "Stonewall" Jackson. The Second Battle of Bull Run was fought on the same field on August 28-30, 1962.

Two months after the Confederate assault on Fort Sumter in Charleston, the American Civil War began in earnest. The First Battle of Bull Run was the North's first attempt at a quick victory by capturing the Confederate capital of Richmond. However, what started promisingly for the Union turned into a rout as northern troops were forced to flee back to Washington. The battle was sobering for the inexperienced troops and commanders of both sides, with high casualties leading to the realization that the war would not be over any time soon.

A little over a year later the Second Battle of Bull Run was fought here on a significantly larger scale but with similar results. The Army of Northern Virginia under the command of Robert E. Lee was attempting to push the Union Army out of Virginia in order to threaten Washington. This time the engagement lasted three days, and despite being outnumbered Lee won the day. His victory cleared the way for a Confederate invasion of Maryland that would only be stopped at Antietam a few months later. Manassas National Battlefield Park was established in 1940 and was listed on the National Register of Historic Places in 1966.

Manassas National Battlefield Park incorporates a large area along Interstate 66 north of the town of Manassas. Some of the highlights include Matthew's Hill, where the first shots were fired at the first battle; Brawner's Farm, where the second battle began; Henry Hill, where Jackson and his brigade made their famous stand; and the Stone Bridge which the Federal armies used during their retreat from both battles. There are museums with exhibits on the battles at the visitor's center and at the old Brawner's Farm building.

---

68. FREDERICKSBURG AND SPOTSYLVANIA NATIONAL MILITARY PARK

1013 Lafayette Boulevard, Fredericksburg, Virginia, 22401

Site Type: Battlefield - Battles of Fredericksburg, Chancellorsville, The Wilderness & Spotsylvania
Conflict: American Civil War
Dates: Battles fought on December 11-15, 1862; April 30-May 6, 1863; May 5-7, 1864 & May 8-21, 1864
Designations: National Register of Historic Places, National Military Park, Virginia State Landmark
Web: www.nps.gov/frsp (official website)

The area around Fredericksburg, Virginia was one of the most hotly contested theaters during the American Civil War. Located almost exactly half-way between the opposing capitals of Washington and Richmond, four major battles occurred within a few miles of each other just west of Fredericksburg. The first two, the Battle of Fredericksburg in 1862 and the Battle of Chancellorsville in 1863, were important victories for the South. The second two, the Battle of the Wilderness and the Battle of Spotsylvania Courthouse in 1864, were both part of the North's ultimately successful drive to capture Richmond.

The Battle of Fredericksburg actually began as part of a Union effort to capture the Confederate capital of Richmond. However, despite significant advantages in numbers the northerners were forced to assault an entrenched Confederate army under the command of Robert E. Lee. The Union army suffered severe casualties and was forced to withdraw. A similar engagement at the Battle of Chancellorsville less than half a year later saw a superior Union army defeated once again by Lee, this time paving the way for the Confederate invasion of Pennsylvania.

The Battles of the Wilderness and Spotsylvania Courthouse were back-to-back engagements that lasted two and a half weeks and pitted Ulysses S. Grant against Lee. Both battles, essentially a delaying action to prevent Grant from capturing Richmond, were inconclusive, with both sides suffering staggering casualties. While the North got the worst of it, the South's casualties were irreplaceable, leaving the Confederacy in a considerably weaker position. Fredericksburg and Spotsylvania National Military Park was established in 1927. It was listed on the National Register of Historic Places in 1966 and designated a Virginia Landmark in 1973.

Fredericksburg and Spotsylvania National Military Park incorporates the bulk of the four battlefields, though they are somewhat separated from each other. Fredericksburg is just south of the city; Spotsylvania two miles further to the southwest; Chancellorsville is three miles to the west; and Wilderness is just west of that. There are several visitor centers for the park, the main one being located in downtown Fredericksburg. Among the sites of particular interest is the house where Stonewall Jackson died at the Battle of Chancellorsville.

69. CEDAR CREEK AND BELLE GROVE NATIONAL HISTORICAL PARK

7712 Main Street, Middletown, Virginia, 22645

Site Type: Battlefield – Battle of Cedar Creek
Conflict: American Civil War
Dates: Battle fought on October 19, 1864
Designations: National Register of Historic Places, National Historic Landmark District, National Historical Park, Virginia State Landmark
Web: www.nps.gov/cebe (official website)

The Battle of Cedar Creek was the final military engagement of the valley campaigns of 1864. This series of battles, fought over several months, was both the last attempt by the Confederacy to capture Washington DC and the Union's effort to control the Shenandoah Valley. The Union victory at Cedar Creek essentially ended the last vestige of hope for a Confederate victory in the war and was a contributing factor to Abraham Lincoln's re-election.

By the spring of 1864 Union forces were on the offensive in every theater of the war. With the Army of the Potomac growing stronger under Ulysses S. Grant and Atlanta threatened with capture, the South undertook one last desperate offensive with the aim of capturing Washington. By August it was clear this was hopeless, and the campaign turned from an advance into a desperate effort to hold the Shenandoah Valley.

On October 19 the Confederate army under Jubal Early made a bold attempt to stop the Union army with a surprise attack at Cedar Creek. The initial attack was a complete success, and the North took heavy casualties. However, the Union rallied under Philip Sheridan, who counterattacked and led the North to a resounding victory, ending any hope for the South. Cedar Creek was named a Virginia State Landmark in 1968, listed on the National Register of Historic Places in 1969 and named a National Historic Park in 2002.

The Cedar Creek and Belle Grove National Historic Park incorporates much of the area on which the Battle of Cedar Creek was fought. Historic markers and monuments note points of interest on the battlefield. Also in the park is the Belle Grove Plantation house, a mansion dating from the 18th century and which Union General Sheridan used as his headquarters during the battle.

---

70. JAMESTOWN SETTLEMENT LIVING MUSEUM

2110 Jamestown Road, Route 31 South, Williamsburg, Virginia, 23185

Site Type: Wooden Stockade Fort
Dates: Originally completed in 1610
Web: www.historyisfun.org/jamestown-settlement/james-fort (official website)

The Jamestown Fort was a wooden stockade that protected the original Jamestown settlement of the early 17th century. It stood for several years, sheltering the colonists during the infamous Starving Time. The location of the original fort can be found at Historic Jamestowne in the Colonial National Historical Park. A replica of the fort now stands about a mile away at the Jamestown Settlement living history museum.

The Jamestown colony was the first permanent English settlement in what later became the Thirteen Colonies, not counting the mysterious abandoned village of Roanoke in North Carolina. The first structure completed was a rudimentary wooden stockade fort inside of which most of the original buildings were later constructed. The fort was briefly abandoned after the Starving Time in 1610, re-occupied shortly thereafter, and then fell into disrepair around 1614.

The Jamestown Settlement Living Museum was established in the 20th century following the successful 350th anniversary celebration of the original settlement in 1957. The main attraction is the reconstructed fort, which encloses a number of other replica buildings. Also part of the park is a recreation of a Powhatan encampment, as well as a replica of one of the ships that brought the first colonists to Virginia.

71. COLONIAL NATIONAL HISTORICAL PARK

1000 Colonial Parkway, Yorktown, Virginia, 23690

Site Type: Battlefield - Siege of Yorktown
Conflict: American Revolution
Dates: Siege fought from September 29-October 19, 1781
Designations: National Register of Historic Places, National Historical Park
Web: www.nps.gov/york (official website)

The Siege of Yorktown was the last major engagement of the American Revolution. After the collapse of their ill-fated Southern campaign, the surviving British army under General Charles Cornwallis was forced to withdraw northwards to Virginia. There they made camp at the port of Yorktown in order to wait for supplies and reinforcements. Instead they found themselves trapped by a combined American-French army as well as a naval blockade. The resultant siege led to the surrender of the largest British army in the colonies.

In the months after the British defeat at the Battle of Cowpens, the remains of the British Southern army headed back towards Virginia. Fighting several smaller engagements along the way, the British eventually made it to the Virginia Peninsula in order to await reinforcements from their garrison in New York. However, Washington got wind of the plan and marched his joint American-French force to Virginia, while a French fleet raced to blockade the peninsula.

By mid-September the British belatedly realized that they were trapped by both land and sea. The ensuing siege lasted for three weeks, and as the British strong points began to fall, the fate of Cornwallis' army was sealed. On October 19th the British force of nearly nine thousand men surrendered. After the siege and fall of Yorktown the British were no longer able to sustain a viable war effort against the Americans and French, and two years later the independence of the colonies was formally recognized under the terms of the Treaty of Paris. Yorktown Battlefield was included as part of the Colonial National Historical Park in 1930.

Colonial National Historical Park actually consists of three major areas on the Virginia Peninsula: Colonial Williamsburg, Jamestown and Yorktown, all connected by the Colonial Parkway. The Yorktown section incorporates most of the Yorktown Battlefield, including preserved earthworks from the British defensive lines. Also on site is the Nelson House where Cornwallis had his final headquarters. The Yorktown Victory Monument stands on the edge of the town of Yorktown close to the battlefield site.

---

72. RICHMOND NATIONAL BATTLEFIELD PARK

500 Tredegar Street, Richmond, Virginia, 23219

Site Type: Battlefield – Peninsula Campaign, Siege of Richmond
Conflict: American Civil War
Dates: Battles fought on March 17-July 1, 1862 and June 15, 1864-April 2, 1865
Designations: National Register of Historic Places, National Historic District, National Battlefield Park, Virginia State Landmark
Web: www.nps.gov/rich (official website)

Richmond National Battlefield Park is one of the more unique of America's military parks. Rather than focus on a single battle, it incorporates multiple sites that involved the Union attempt to capture Richmond and the Confederate effort to defend the city over the course of an extended period of time. It includes both locations where fighting took place as well as the Confederacy's largest ironworks and hospital.

The city of Richmond was designated the capital of the Confederacy soon after Virginia's secession from the United States. As such it became one of the prime strategic targets of the Union army. Efforts to capture Richmond took place as early as the spring of 1862, when a Union army under the command of George McClellan attempted but failed to take Richmond in a series of engagements now referred to as the Peninsula Campaign.

Two years later Ulysses S. Grant led a second attempt to take Richmond in what is now known as the Overland Campaign. The initial effort to assault the city was a failure, so Grant instead spent nearly a year laying siege to the nearby town of Petersburg. Shortly after Petersburg was taken Richmond fell into Union hands in April of 1865. The Richmond National Battlefield Park was established in 1936, listed on the National Register of Historic Places in 1966, and named a Virginia State Landmark in 1973.

Richmond National Battlefield Park is spread out over more than a dozen locations with multiple visitor centers. Most of the sites are related to the Peninsula Campaign and include several places where the Confederates mounted their defenses. There are also sites from the Overland Campaign, including the Cold Harbor Battlefield, the site of Robert E. Lee's last victory. Within the city of Richmond are two places of particular interest: the Tredegar Iron Works, where most of the South's artillery was produced; and the Chimborazo Hospital, where thousands of Confederate wounded were cared for during the war.

---

73. PETERSBURG NATIONAL BATTLEFIELD

5001 Siege Road, Prince George, Virginia, 23803

Site Type: Battlefield - Siege of Petersburg
Conflict: American Civil War
Dates: Siege fought from June 9, 1864-March 25, 1865
Designations: National Register of Historic Places
Web: www.nps.gov/pete (official website)

The Siege of Petersburg was the longest military engagement of the Civil War. Ending with the capture of Petersburg by Union forces in the spring of 1865, the city's fall led directly to the capture of the Confederate capital of Richmond and the surrender of the Army of Northern Virginia a few weeks later. The siege was perhaps most famous for the Battle of the Crater, an ill-fated attempt to undermine the Confederate defenses that led to more than five thousand casualties in just a few hours.

In 1864 Ulysses S. Grant launched his campaign to capture the city of Petersburg. Petersburg protected the southern approaches to Richmond, and Grant figured correctly that its capture would lead directly to the fall of the capital. The Army of Northern Virginia under Robert E. Lee established a massive series of earthworks to defend Petersburg and what resulted was more than nine months of trench warfare. The Union Army made several attempts to overcome the defenses throughout 1864, including the Battle of the Crater.

In this infamous engagement which took place on July 30, a tunnel was dug and massive amounts of explosives planted beneath the Confederate defense. A huge gap was blown in the defensive line and in rushed the Union army, only to find themselves trapped in a crater which they could not get out of. Thousands were killed in the ensuing fighting, and the Union army failed to exploit the breach. The Confederate Army was finally forced to abandon the city in March of 1865. Petersburg National Battlefield was listed on the National Register of Historic Places in 1966.

Petersburg National Battlefield consists of several independent sites of interest around the city of Petersburg. The main site at Eastern Front has the visitor center. This part of the park is home to Fort Stedman, a Confederate strongpoint, and the site of the Crater. The entrance to the tunnel constructed by the Union army has been restored. Many of the dead from the siege are buried at nearby Poplar Grove National Cemetery.

---

74. UNITED STATES ARMY WOMEN'S MUSEUM

2100 A Avenue, Fort Lee, Virginia, 23801

Site Type: Museum
Dates: Opened in 2001
Web: www.awm.lee.army.mil (official website)

The United States Army Women's Museum is the largest facility of its kind exclusively honoring the role of women serving in the American military. Originally founded at Fort McClellan, Alabama in 1955, it was relocated to a new facility at Fort Lee in 2001. As of the time of this writing the museum had recently undergone a major overhaul, with new exhibits and collections added in 2018.

Women have served in the United States Army since the American Revolution. Prior to the 20th century, all such service was done in secret, with many women disguised as men fighting in various conflicts. Female enlistment was finally allowed in the early 1900s, with recruiting of nurses prior to World War I, and greatly expanded with the creation of the Women's Army Corps during World War II.

Barriers continued to come down in 1976, when the first female cadets were admitted to West Point; in 1978, when the WACs were integrated with all-male units; and in the 2000s, when women were increasingly allowed to participate in combat. As of the time of this writing, women made up approximately 16% of Army active service personnel.

The United States Army Women's Museum is a large facility located at the Fort Lee army base, which is home to a number of army commands and schools. Exhibits are organized in a series of roughly chronological galleries, from origins of women in the military to service in the 21st century. One of the museum highlights is the exhibit on women who served in World War II, which provided the springboard for the massive expansion of women in the military in the modern era.

---

75. USS WISCONSIN

1 Waterside Drive, Norfolk, Virginia, 23510

Site Type: Battleship
Conflicts: World War II, Korean War
Dates: Commissioned in 1944
Designations: National Register of Historic Places, Virginia Historic Landmark
Web: https://nauticus.org/battleship-wisconsin (official website)

The USS Wisconsin (BB-64), or *Wisky*, was the very last battleship in American history to be commissioned. It was also one of the very last to be decommissioned. The Wisconsin saw active service during World War II and the Korean War, and was one of two battleships, along with the Missouri, to see combat during the Gulf War. The Wisconsin received its final decommission in 1991. It was established as a museum ship in 2001.

The Wisconsin was completed in 1943 and entered active service in 1944. It spent the last year of the war in the Pacific theater and supported the American landings at both Iwo Jima and Okinawa. The Wisconsin spent the last days of the war shelling targets on the Japanese mainland. After World War II she served in a number of capacities including as a training vessel.

Over the course of nearly half a century the Wisconsin was decommissioned and recommissioned twice. In 1952 she went into action again, this time to support ground combat in Korea. On February 28, 1991 the Wisconsin fired on Iraqi targets during the Gulf War. This was the last time any battleship anywhere in the world fired its guns in a combat capacity.

The USS Wisconsin is permanently moored in Norfolk, Virginia. After its final decommissioning it was one of two battleships that were required to stay in a state of readiness as a reserve vessel. Although this requirement ended in 2009, because of it the Wisconsin is arguably in the finest surviving condition of any United States battleship still afloat.

---

76. MACARTHUR MEMORIAL MUSEUM

198 Bank Street, Norfolk, Virginia, 23510

Site Type: Museum
Dates: Opened in 1964
Web: www.macarthurmemorial.org (official website)

The MacArthur Memorial Museum is dedicated to the life of one of America's most storied and controversial military commanders,

Douglas MacArthur. MacArthur's career spanned both World Wars and the Korean conflict. As one of the highest ranking American officers during World War II he was instrumental in planning the campaign in the Pacific Theater and the defeat of Imperial Japan.

Douglas MacArthur began his career as a top performing cadet at West Point. He achieved the rank of general during World War I, before he turned forty. Between the wars he became a major general, participated in the controversial crushing of the Bonus Army in 1932, and endured a rocky relationship with President Roosevelt. He commanded the defense of the Philippines when it fell to the Japanese in 1942, and led the American forces to its recapture in 1944. He retired in 1951 as one of the most popular figures in American life.

The MacArthur Memorial Museum is located on MacArthur Square in the heart of Norfolk. The museum is full of galleries with a large collection of military and personal memorabilia associated with MacArthur's life. There are also exhibits on those who served under him. Douglas MacArthur and his wife are entombed in the rotunda of the museum.

---

77. APPOMATTOX COURT HOUSE NATIONAL HISTORICAL PARK

111 National Park Drive, Appomattox, Virginia, 24522

Site Type: Battlefield – Battle of Appomattox Court House
Conflict: American Civil War
Date: Battle fought on April 9, 1865
Designations: National Register of Historic Places, National Historic District, Virginia State Landmark
Web: www.nps.gov/apco (official website)

The Battle of Appomattox Court House was the last military engagement of consequence during the Civil War. It was here on April 9 that the bedraggled Army of Northern Virginia, outnumbered by more than five to one, was finally cornered by the Union army. After a brief last stand, the South's greatest general surrendered its largest remaining army, effectively ending the war.

Following the Confederate loss of its capital city of Richmond, Robert E. Lee rallied what was left of his forces, which numbered less than thirty thousand men, and marched with them into central Virginia intent on linking up with the Army of Tennessee. He was pursued by the much larger Army of the Potomac under Ulysses S. Grant. Throughout the first week of April the Union cavalry wreaked havoc on Lee's army, disrupting the Confederate movements and seizing or destroying their supply trains.

By April 9 the Union forces were closing in all around the Confederates. Lee attempted a last desperate assault in order to break out of the trap but the effort was doomed from the start. The South took heavy losses, and before long Lee determined there was no other option but for him to surrender. The official surrender ceremony took place on April 12. Appomattox Court House was listed on the National Register of Historic Places in 1966 and named a Virginia State Landmark in 1971.

Appomattox Court House National Historical Park encompasses much of the area where the fighting of the battle took place. It includes historic buildings, but these are for the most part recreations. The McLean House, where the surrender took place, is also a recreation from the 1940s. Despite this the room where Grant met with Lee to sign the surrender papers is the most popular visitor destination in the park.

---

78. NATIONAL CIVIL WAR CHAPLAINS MUSEUM

1971 University Boulevard, Lynchburg, Virginia, 24515

Site Type: Museum
Conflict: American Civil War
Dates: Opened in 2005
Web: www.liberty.edu/arts-sciences/history/national-civil-war-chaplains-museum (official website)

The National Civil War Chaplains Museum is a gallery on the campus of Liberty University dedicated to the clergymen of all faiths who served during the Civil War. Representing those who ministered to the troops of both the Union and the Confederacy, it is the only such museum in the United States, with an exceptional collection of military-related religious artifacts.

Chaplaincy has been a tradition of the American military since the founding of the United States. In no American war did chaplains play a more important role than in the Civil War, which pitted brother against brother and resulted in more American deaths than any other conflict in history. Nearly four thousand clergymen of many faiths served during the conflict.

The National Civil War Chaplain's Museum is home to a number of exhibits, including the role of the clergy in the military, the influence of religion at the time of the Civil War and the United States Christian Commission, a home front organization that provided care and comfort to soldiers. While the focus is on Christianity, there are exhibits on Jewish chaplains as well.

---

79. VIRGINIA MILITARY INSTITUTE

319 Letcher Avenue, Lexington, Virginia, 24450

Site Type: Military College
Dates: Opened in 1839
Web: www.vmi.edu (official website)

The Virginia Military Institute is one of only a handful of senior military colleges, along with the academies, in the United States, and arguably the most historic. Nicknamed the West Point of the South, VMI has turned out some of the most important American army officers outside of the United States Military Academy. Unlike most of the other military colleges, the Virginia Military Institute is exclusively for cadets.

The Virginia Military institute was founded in 1839 in answer to a need for better trained leadership for the Virginia militia. Many of its early graduates went on to serve during the Civil War, mostly on the Confederate side, though some did fight for the Union. One of the institute's first professors, Thomas "Stonewall" Jackson, was among the most renowned Southern commanders of the war. Over two hundred cadets fought at the Battle of Newmarket in 1864.

VMI also turned out many officers that served in the World Wars and other conflicts of the 20th century. Although George Patton attended the institute, he did not graduate from here. However, George Marshall, one of only handful of men ever to achieve the rank of 5-star general, did graduate from VMI in 1901. Marshall had an illustrious postwar career that included serving as Secretary of State and receiving the Nobel Peace Prize.

The Virginia Military Institute campus has a number of sites worth seeing, some of which can be visited by tour. The VMI Museum houses a variety of artifacts, mostly related to the faculty and cadets who attended the college. There is also the Virginia Museum of the Civil War and the Marshall Museum, dedicated to the life of VMI's most famous graduate. Just off campus is the Stonewall Jackson House, which is separately listed on the National Register of Historic Places.

---

80. HARPER'S FERRY NATIONAL HISTORICAL PARK

171 Shoreline Drive, Harper's Ferry, West Virginia, 25425

Site Type: Battlefield – Raid on Harper's Ferry
Conflict: Raid on Harper's Ferry
Dates: Raid took place on October 18, 1859
Designations: National Historical Park
Web: www.nps.gov/hafe (official website)

Harper's Ferry National Historical Park is home to several locations associated with John Brown's famous ill-fated 1859 raid on the local armory. The site of one of America's most important military arsenals, Brown, an ardent abolitionist, hoped that the raid would help to trigger and arm a slave revolt in Northern Virginia. Although a failure, the raid galvanized Northern anti-slavery sentiment and served as an opening act for the American Civil War.

John Brown is one of the most controversial figures in American history. A Northerner who fought his entire life to see the end of slavery, Brown was among those who believed that only violence would put an end to the hated institution. After fighting against slavery in Bloody Kansas he moved back east where he conceived a plan to instigate a slave revolt in Virginia. Supported by other anti-slavery figures including Harriet Tubman, he organized a small force of about two dozen men and marched on Harper's Ferry.

On October 16, 1859 Brown's men seized the Harper's Ferry arsenal. However, the local militia soon turned out, preventing the raiders from getting away. Word reached Washington, DC, and a small force led by future Confederate general J.E.B. Stuart was sent out to bring the situation under control. He reached Harper's Ferry on October 18, quickly killed or captured the entire raiding force, and seized John Brown, who was subsequently hanged for treason. Harper's Ferry was declared a National Historical Park in 1963.

Harper's Ferry National Historic Park incorporates the Harper's Ferry Historic District, with a number of sites related to the raid. The most important and popular of these is referred to as John Brown's Fort. Actually an auxiliary support building for the armory, it was here that Brown and his men made their final stand. Also in the historic district is the John Brown Wax Museum, with more information on his life and the raid.

MIDWEST

81. PERRYVILLE BATTLEFIELD STATE HISTORIC SITE

1825 Battlefield Road, Perryville, Kentucky, 40468

Site Type: Battlefield – Battle of Perryville
Conflict: American Civil War
Dates: Battle fought on October 8, 1862
Designations: National Register of Historic Places, National Historic Landmark, Kentucky State Historic Site
Web: www.perryvillebattlefield.org (official website)

The Battle of Perryville was the largest military engagement ever fought in the state of Kentucky. The culmination of one of the lesser known campaigns of the Civil War, this incredibly bloody battle left the Union in control of the state and denied the Confederacy access to the strategically critical Ohio River Valley for much of the war.

At the outset of the Civil War, Kentucky, one of the border states, initially took a neutral stance towards the conflict. However, the Confederacy moved forces into Kentucky in the summer of 1861. This threatened the Union along its entire Western flank, and a major campaign was undertaken to secure Kentucky for the North. The buildup of forces and skirmishing went on through most of 1862. Things came to a head in October when the main Confederate army came up against part of the Union army at Perryville.

Because most of the Union force was not present at the battle, the Confederacy was able to drive them from the field. Casualties were extremely high, approximately twenty percent for each side. Despite the Confederate victory, the rebels were forced to retreat to Tennessee when the rest of the Union Army showed up. Perryville Battlefield was established as a Kentucky State Historic Site in 1954 and listed on the National Register of Historic Places in 1966.

Perryville Battlefield State Historic Site incorporates a large part of the battle site. The main area of interest is near the visitor's center, where monuments stand to both the Union and Confederate forces. The battlefield is marked with signage tracing the progress of the engagement.

---

82. FORT KNOX & GENERAL GEORGE PATTON MUSEUM

4554 Fayette Avenue, Fort Knox, Kentucky, 40121

Site Type: Military Base, Museum
Dates: Base established in 1918
Web: www.generalpatton.org (official website)

Fort Knox is one of the best known military bases in American history, the main reason for this being that it is the repository for much of the gold reserves of the United States. In actuality the gold reserves are kept next door at the United States Bullion Depository. But sprawling Fort Knox has an impressive history in its own right, and is home to a number of major military units and commands. It is also home to the General George Patton Museum.

The first military facility in the area was Fort Duffield. Completed in 1861, the fort only played a very minor role in the American Civil War, and in fact was largely abandoned throughout most of the conflict. The area around Duffield started to be used for training in the early 20th century, and in 1918 Fort Knox was established. During the interwar years Fort Knox became a major training center for mechanized units, and served as an important military base during World War II.

Over the years Fort Knox has been home to many military commands, with a number of changes over time. Several divisions are based here, as well as major logistical units, Army reserve aviation, military intelligence, and, most recently, the Army Recruiting Command.

The General George Patton Museum was opened in 1948 at Fort Knox due to its importance in the role of mechanized warfare during World War II. The museum formerly had a large collection of tanks and mechanized vehicles, but many of these have since been relocated to other facilities. However, it does have a substantial collection of exhibits related to its namesake, General Patton, including personal artifacts from his life.

---

83. NATIONAL MUSEUM OF THE UNITED STATES AIR FORCE

1100 Spaatz Street, Dayton, Ohio, 45433

Site Type: Museum
Dates: Opened in 1954
Web: www.nationalmuseum.af.mil (official website)

The National Museum of the United States Air Force is the oldest and largest military aviation museum in the world. Located at Wright-Patterson Air Force Base, it is actually older than the Smithsonian Air and Space Museum in Washington. Built around a private collection that started a few years after the end of the First World War, the museum has been collecting aircraft and related artifacts throughout the entire history of military air power.

Wright-Patterson Air Force Base is one of America's most important military installations. Established in 1917 as Wilbur Wright Field, it is associated with the Huffman Prairie Flying Field where the Wright Brothers tested some of their airplanes. The original museum collection began in 1923 with a few pieces assembled by the base's engineers. It moved to its current location in the 1950s and subsequently expanded to four hangars. It was designated the National Museum of the United States Air Force in 2004.

The National Museum of the United States Air Force is home to over three hundred airplanes and thousands of other aviation-related artifacts. Virtually every class of American military aircraft is represented, plus many from foreign countries. Highlights include a biplane used by the Lafayette Escadrille; the Bockscar, which dropped the atomic bomb over Nagasaki, Japan; the Memphis Belle; and the plane on which Lyndon Johnson was sworn into office after the Kennedy assassination. Also on display are uniforms worn by famous servicemen, including Billy Mitchell, Jimmy Stewart and Ronald Reagan.

---

84. CHARLES YOUNG BUFFALO SOLDIERS NATIONAL MONUMENT

1120 U.S. Route 42, Wilberforce, Ohio, 45384

Site Type: Museum
Dates: Opened in 2013
Designations: National Register of Historic Places, National Historic Landmark, National Monument
Web: www.nps.gov/chyo (official website)

The Charles Young Buffalo Soldiers Monument is named for Charles Young, a soldier in the United States Army who had what was arguably the most distinguished military career for any African American prior to World War II. Born a child of slaves in Kentucky in 1864, he ultimately rose to the rank of Lieutenant Colonel, and might have risen higher had it not been for the prejudices in the military that were common at the time.

Young was among the first African Americans to attend West Point. His main service after graduation was in the cavalry, notably with the famous Buffalo Soldiers. He fought in the Spanish-American War and for a time served in a senior position with the National Park Service. In his later years of service he achieved the rank of Lieutenant Colonel.

Young was forced into retirement when the United States entered World War I due to concerns among white officers that he would be promoted to the rank of general. He died in 1922 and is buried at Arlington National Cemetery. The Charles Young Buffalo Soldiers National Monument was listed on the National Register of Historic Places and named a National Historic Landmark in 1974.

The Charles Young Buffalo Soldiers National Monument is

located at Young's former home in Ohio. The house, which dates from the 1830s, may have been used to hide escaped slaves along the Underground Railroad. It is now operated as a museum, with exhibits on his life and on the Buffalo Soldiers cavalry units with which he served for much of his career.

---

85. PERRY'S VICTORY AND INTERNATIONAL PEACE MEMORIAL

93 Delaware Avenue, Put-In-Bay, Ohio, 43456

Site Type: Monument – Battle of Lake Erie
Conflict: War of 1812
Dates: Battle fought on September 10, 1813; Monument Dedicated in 1931
Designations: National Register of Historic Places
Web: www.nps.gov/pevi (official website)

The Battle of Lake Erie was one of the largest naval engagements of the War of 1812, the largest ever fought on the Great Lakes, and one of the largest in history fought on a body of freshwater. One of the decisive battles of the conflict, it resulted in American control of the Midwest and made Oliver Perry a household name. The monument to the battle was built on the shore at one of the closest points to where the fighting occurred.

The northwestern theater of the War of 1812 between the United States and Great Britain revolved around control of the Great Lakes, primarily Erie. The British seized control of Erie early on during the conflict, threatening the entire American frontier in the Midwest. Throughout 1812 and early 1813 both sides scrambled to strengthen their naval forces on the lake, with the Americans ultimately gaining the upper hand.

When the two fleets met on September 10, the fighting was ferocious. Both sides suffered significant casualties. However, the larger American fleet under the command of Oliver Perry won the battle decisively with the capture of the entire British squadron. In 1915 a massive victory column was erected at Put-In-Bay to commemorate the battle. It was listed on the National Register of Historic Places in 1966.

Perry's Victory and International Peace Memorial is a dual-purpose monument: first, it celebrates Oliver Perry's decisive victory at the Battle of Lake Erie. Second, it celebrates the long-standing peace, now over two centuries old, between the United States, Canada and Great Britain. The column, at 352 feet in height, is the third tallest commemorative monument in the United States. The top is accessible by elevator. Several sailors killed at the battle are buried near the base of the column.

---

86. FALLEN TIMBERS BATTLEFIELD AND FORT MIAMIS NATIONAL HISTORIC SITE

North Jerome Road, Maumee, Ohio, 43537

Site Type: Battlefield – Battle of Fallen Timbers
Conflict: Early American Indian Wars
Dates: Battle fought on August 20, 1794
Designations: National Register of Historic Places, National Historic Site, National Historic Landmark
Web: www.nps.gov/fati (official website)

The Battle of Fallen Timbers was the largest battle fought on American soil between the end of the American Revolution and the beginning of the War of 1812. The decisive engagement of the Northwest Indian War, Fallen Timbers was part of the early effort to expand the American frontier into the Midwest at the expense of local tribes. It was also a proxy engagement of the ongoing conflict between the Americans and British, with the latter supporting the Native Americans.

After the Treaty of Paris formally ended the American Revolution in 1783, the new country almost immediately began to expand westward. The first major national effort in this arena was to secure the Ohio territory in what later became known as the Northwest Indian War. For the better part of a decade a large confederation of local tribes, backed by the British, was mostly successful in keeping American colonists out of Ohio.

In 1794 a better organized, trained and equipped force under the leadership of Anthony Wayne was dispatched to the frontier. The two armies met at Fallen Timbers in August. However the British decided not to participate, and without their support the Native defenders were quickly routed. This opened the way to the American annexation of territory in Ohio and set the stage for further expansion westward. The battlefield was named a National Historic Landmark in 1960, listed on the National Register of Historic Places in 1966, and designated a National Historic Site in 1999.

The Fallen Timbers Battlefield National Historic Site consists of several areas, including the location where the battle was incorrectly thought to have taken place, a new location where the battle actually took place, and the surviving earthworks of the British Fort Miamis. The centerpiece of the park is the Battle of Fallen Timbers Monument, located at the original incorrect battle site. The monument features statues of Anthony Wayne, a frontier militiaman and a Native American scout, honoring all of those who fought at the battle.

---

87. POLAR BEAR EXPEDITION MEMORIAL

621 West Long Lake Road, Troy, Michigan, 48098

Site Type: Monument – Polar Bear Expedition
Conflict: Russian Civil War
Dates: Dedicated in 1930
Web: www.whitechapelcemetery.com (official website)

The Polar Bear Expedition Memorial at White Chapel Memorial Cemetery in Michigan commemorates one of the least known military operations in American history: the United States effort to assist the White Army during the Russian Civil War. Detached to Russia even before the First World War had ended, these troops fought to prevent the Communist takeover of Russia. A number of those who died during this operation are buried in this cemetery.

Following the overthrow of the Romanov dynasty in 1917, Russia withdrew from World War I and was plunged into Civil War. There were a number of forces in play, but the two main groups were the largely democratic White Army and Communist Red Army. Although Russia was no longer fighting Germany and Austria, her Allies came to the aid of the White Army. Forces from the United States, the British Empire, France, Japan and Czechoslovakia all came to help the Whites.

The United States organized the American Expeditionary Force in North Russia, better known as the Polar Bear Expedition. Consisting of about five thousand troops, the AEF was sent to Archangel in an effort to bolster local White Russian forces. Unfortunately the campaign was unsuccessful, and without sufficient local support the Americans were forced to withdraw their forces in 1919.

The Polar Bear Expedition Memorial can be found in the White Chapel Cemetery in Troy, Michigan. The soldiers of this expedition suffered over two hundred casualties in Russia, more than fifty of whom are now buried in this cemetery. The Polar Bear Expedition Memorial was dedicated in 1930.

---

88. RIVER RAISIN NATIONAL BATTLEFIELD PARK

1403 East Elm Street, Monroe, Michigan, 48162

Site Type: Battlefield – Battle of Frenchtown
Conflict: War of 1812
Dates: Battle fought on January 18-23, 1813
Designations: National Register of Historic Places, National Battlefield Park, Michigan State Historic Site
Web: www.nps.gov/rira (official website)

The Battle of Frenchtown was the worst defeat for the American army during the War of 1812 and one of its biggest losses in the pre-Civil War era. Actually a series of engagements that took place in January of 1813, the fighting originally went in favor of the Americans, but ended a few days later in a catastrophic defeat and massacre. This victory helped to secure Michigan for the British for a short period, at least until Perry's victory at the Battle of Lake Erie a few months later.

The battle to control the American northwestern frontier was one of the major factors leading up to the War of 1812, and initially things did not go well for the United States. A stronger naval presence on Lake Erie, combined with assistance from Native American allies, led to early British victories in Ohio and Michigan. In August of 1812 the Americans were forced to surrender Fort Detroit, leaving settlements in Michigan defenseless. An expeditionary force under future president William Henry Harrison was sent to recover the fort.

In the ensuing campaign Harrison unwisely divided his forces. Part of the British army came up against one of the American columns on January 18 and was driven off after a skirmish. A few days later the two sides clashed again, only this time the British had their full force and the Americans were routed. Nearly half were killed and the balance captured. This defeat seriously curtailed American campaigning in the region for the better part of the next year. The battlefield was named a Michigan State Historic Site in 1956, added to the National Register of Historic Places in 1982, and designated a National Battlefield Park in 2009.

River Raisin National Battlefield Park only includes part of the battlefield site due to partial private ownership of the land. As of the time of this writing the park was still being developed, and not all of it is currently accessible to visitors. However there is a visitor center with exhibits on the battle, and a few markers noting where pivotal events of the fighting took place.

---

89. YANKEE AIR MUSEUM

47884 D Street, Belleville, Michigan, 48111

Site Type: Museum
Dates: Opened in 1981
Web: http://yankeeairmuseum.org (official website)

The Yankee Air Museum is a Smithsonian affiliated aviation museum that is home to a number of military aircraft and related artifacts. Part of the museum's collection, which has expanded significantly since the facility opened, is housed in a former bomber factory. Several of the planes here are maintained in flying condition and rides may be available, making this one of the few places where visitors can experience flight in a genuine World War II aircraft.

The Yankee Air Museum first opened in 1981, and the current facility has been in operation since 2010. In its four decades the museum has acquired a considerable collection of aircraft and gliders, mostly from World War II and the early Cold War periods. There are also exhibits on the history of the Ford Willow Run bomber factory and Rosie the Riveter, aviation technology, and air combat during World War II and the Vietnam War.

The Yankee Air Museum's main attractions are undoubtedly the collection of serviceable military aircraft, which fly with passengers during the summer months. This collection includes a Waco Biplane, a C-47 Skytrain, a B-25 Mitchell, B-17 Flying Fortress and a UH-1 Huey helicopter. All of these aircraft offer short flights out of Willow Run Airport for an extra charge. It is highly recommended that those who wish to fly check ahead for availability and schedules.

---

90. INDIANA WORLD WAR MEMORIAL PLAZA

55 East Michigan Street, Indianapolis, Indiana, 46204

Site Type: Monument – Indiana War Veterans, Museum
Dates: Dedicated in 1924
Designations: National Register of Historic Places, National Historic District
Web: www.in.gov/iwm/2333.htm (official website)

The Indiana World War Memorial Plaza is home to one of the largest military monument complexes in the United States outside of Washington DC. Originating as the site of the Soldiers and Sailors Monument, the plaza was greatly expanded at the beginning of the 20th century to include a massive veteran's memorial, the Indiana World War Memorial Military Museum and the national headquarters of the American Legion.

The plaza's first structure was the Indiana State Soldiers and Sailors Monument. Completed in 1902, this immense monument towers 284 feet in height over the surrounding square. Originally dedicated to Indiana veterans of the Civil War, it was later expanded to honor those who had served in the major American conflicts from the Revolution through the Spanish-American War.

In 1919 Indianapolis was chosen to be the home of the national Headquarters of the American Legion, which had been formed by World War I veterans. The plaza was expanded to incorporate a new space, Cenotaph Square, where the American Legion building now stands. Memorials outside of the American Legion honor those who served in World War II, Korea and Vietnam.

During the 1920s construction began on the massive War Memorial building. Modeled after the Mausoleum of Halicarnassus, it was not completed until the 1960s. In addition to honoring those who served in World War I, it houses the Indiana World War Memorial Military Museum. The Indiana World War Memorial Plaza was listed on the National Register of Historic Places in 1989 and named a National Historic Landmark District in 1994.

The Indiana Soldiers' and Sailors' Monument is the centerpiece of the Indiana World War Memorial Plaza. The defining landmark of downtown Indianapolis, it towers over the streets below. Statues of local heroes grace various levels of the monument, with an immense statue of Victory crowning the top. A small Civil War museum is located at the base of the structure. An observation deck at the top is accessible by elevator or stairs.

---

91. USS INDIANAPOLIS NATIONAL MEMORIAL

Canal Walk, Indianapolis, Indiana, 46202

Site Type: Monument – USS Indianapolis
Conflict: World War II
Dates: Dedicated in 1995
Web: www.in.gov/iwm/2328.htm (official website)

The USS Indianapolis is one of the most famous ships in the history of the United States Navy. In service in the Pacific throughout most of World War II, it transported the pieces of the atomic bomb known as *Little Boy* to Tinian Island. A few days later its sinking resulted in the death of most of its crew. The story was famously recounted in the 1975 movie *Jaws*.

In 1945 the Indianapolis crossed the Pacific Ocean in a then-record time of ten days. Unknown to most of her crew the ship was carrying the components of the atomic bomb that would later be dropped on Hiroshima. The bomb was delivered in utmost secrecy to the bomber base at Tinian Island on July 26. Four days later, as the Indianapolis sailed towards the Philippines, she was sunk by a Japanese submarine.

The Indianapolis went down fast, taking several hundred of the crew members with her. The balance, who escaped into the ocean, were largely without life boats or vests. Due to an oversight the navy was unaware of the sinking for several days, by which time most of the survivors had perished. Exposure, fatigue and shark attacks took hundreds of lives, and barely one fourth of the nearly twelve hundred crew members were rescued.

The USS Indianapolis Memorial was erected in its namesake city of Indianapolis in 1995. Located on the city's canal walk, it features a large granite structure engraved with the outline of the ship and the names of all of those who served and died with her. The memorial was funded in large part by survivors of the disaster.

---

92. TIPPECANOE BATTLEFIELD PARK

200 Battle Ground Avenue, Battle Ground, Indiana, 47920

Site Type: Battlefield – Battle of Tippecanoe
Conflict: Early American Indian Wars
Dates: Battle fought on November 7, 1811
Designations: National Register of Historic Places, National Historic Landmark
Web: www.tippecanoehistory.org/our-places/Tippecanoe-battlefield-museum (official website)

The Battle of Tippecanoe was one of the definitive military engagements in the early years of the Native American Indian Wars. Fought in 1811, a force of about one thousand Americans under the leadership of William Henry Harrison took on about six hundred warriors of the Tecumseh Confederacy. An American victory of sorts, the battle contributed to the tension between the United States and Great Britain that led the next year to the War of 1812.

After the American Revolution settlers began moving west into what is now Indiana. Under the leadership of the popular chief Tecumseh, and at the encouragement and with the support of the British, Native American tribes began to organize a resistance to the settlements. Tecumseh's Confederacy actively opposed large-scale settlement from around 1805 to the end of the War of 1812. The most famous clash, at least from an American standpoint, was the Battle of Tippecanoe.

In order to discourage Tecumseh's army, Harrison led a mixed force of regulars and militia to destroy the Native American stronghold of

Prophetstown. Although Tecumseh was not present at the time, his brother led the warriors in a surprise attack against the Americans. After two hours of vicious fighting, with roughly equal casualties, Tecumseh's army was forced to withdraw due to ammunition shortages. Harrison's force subsequently destroyed Prophetstown, and the future president declared victory. Tippecanoe Battlefield Park was named a National Historic Landmark in 1960 and listed on the National Register of Historic Places in 1966.

Tippecanoe Battlefield Park incorporates much of the original battle site as well as a museum. An obelisk monument commemorates the encounter. While the battle was one of the more even engagements of the Native American Indian Wars, in hindsight it is not regarded as a glorious American victory, and Tecumseh is now highly regarded by many as a national hero.

---

93. NATIONAL VETERAN'S ART MUSEUM

4041 North Milwaukee Avenue, Chicago, Illinois, 60641

Site Type: Museum
Dates: Opened in 1980
Web: www.nvam.org (official website)

The National Veteran's Art Museum, originally known as the National Vietnam Veteran's Art Museum, is the only museum of its kind in the United States. Kept on display here are thousands of pieces of art produced by United States veterans. Artworks date from many American wars, with a focus on Vietnam and other conflicts of the 20th century.

The museum was started in 1981 as a travelling exhibit of art produced by veterans of the Vietnam conflict. With the assistance of then mayor Richard Daley, the exhibit found a permanent home in Chicago in 1995. Since then the museum collection has grown to include works from wars beyond Vietnam, with donations from veterans all over the country.

The exhibits at the National veteran's Art Museum change periodically. At the time of this writing there was a portrait gallery featuring paintings and drawings done by veterans as well as portraits of veterans done by artists. Particularly poignant is the exhibit *Above and Beyond*, a memorial consisting of nearly sixty thousand dog tags representing all those who died in the Vietnam War.

---

94. FORT DE CHARTRES STATE HISTORIC SITE

1350 State Route 155, Prairie Du Rocher, Illinois, 62277

Site Type: Masonry Fort
Dates: Originally completed in 1760
Designations: National Register of Historic Places, National Historic Landmark, Illinois State Historic Site
Web: www.fortdechartres.us (official website)

Fort Du Chartres was one of the strongest and most important military fortifications erected along the Mississippi River during the Colonial era. Built by the French to protect their trade interests in the area, the stone fort was only in use for barely more than a decade before being abandoned. Nevertheless it is still considered to be one of the most important military installations of its era in the Midwest.

The first fort to be built on the current site of Fort Du Chartres was a wooden stockade constructed in 1720. The original fort did not last long and was soon made unusable by river flooding. A successor fort, also of wood, was constructed, but by the 1740s was also becoming unusable. This was finally replaced by the stone fort in 1760.

Fort Du Chartres changed hands only three years later, when Britain took possession of the territory following the French and Indian War. They in turn abandoned it in 1772, just a few years before the American Revolution broke out. It was not until the 1930s that the site of the fort, by then almost a total ruin, was restored by the state of Illinois. Fort Du Chartres was named a National Historic Landmark in 1960 and listed on the National Register of Historic Places in 1966.

Fort Du Chartres appears to rise from the Mississippi plain like some relic of the Late Middle Ages. It is, with perhaps the exceptions of the Castillo de San Marcos in Florida and Fort Ticonderoga in New York, the most impressive Colonial-era fort standing in America. However this is somewhat deceptive as much of the fort is a reconstruction. The oldest surviving building in the fort is the powder magazine.

---

95. USS COBIA & WISCONSIN MARITIME MUSEUM

75 Maritime Drive, Manitowoc, Wisconsin, 54220

Site Type: Submarine, Museum
Conflict: World War II
Dates: Commissioned in 1944
Web: www.wisconsinmaritime.org (official website)

The Wisconsin Maritime Museum was established in the 1960s to honor the city of Manitowoc's role in the manufacturing of submarines during the Second World War. Not exclusively a military museum, there are a variety of exhibits on the maritime history of the Great Lakes. However, the highlight of the museum's collection is the USS Cobia, permanently moored right next door.

The USS Cobia was one of the many submarines dispatched to the Pacific Theater during World War II. Active for the last year and a half of the war, the Cobia sank nearly seventeen thousand tons of Japanese ships. After the war the Cobia was decommissioned, then recommissioned as a training ship. She received her final decommission in 1954. The Cobia became part of the Wisconsin Maritime Museum in 1986.

The USS Cobia has been largely restored and is one of the best preserved World War II submarines in the United States. Almost the entire ship is open to visitors. In addition, the museum houses an exhibit on the Cobia, focusing on life in a submarine and finishing with a simulated attack by a depth charge. Another exhibit features the history of Wisconsin's naval shipbuilding history.

SOUTH ATLANTIC

96. USS NORTH CAROLINA

1 Battleship Road, Wilmington, North Carolina, 28401

Site Type: Battleship
Conflict: World War II
Dates: Commissioned in 1940
Designations: National Register of Historic Places, National Historic Landmark
Web: www.battleshipnc.com (official website)

The USS North Carolina (BB-55), or the *Showboat*, was one of the last battleships commissioned prior to America's entry into World War II in 1941. Originally scheduled to be sent to Pearl Harbor with the rest of the Pacific fleet, it narrowly avoided that disaster by being assigned to patrol duty on the East Coast instead. It was rushed to the Pacific theater to relieve the devastated fleet at Pearl Harbor in 1942. The North Carolina's service ended almost immediately after the war, and she was permanently decommissioned in 1947. It was established as a museum ship in 1986.

The North Carolina began its career during World War II as an escort ship in the North Atlantic to protect Allied transports and cargo from marauding German surface ships. She was redeployed to the Pacific in 1942, arriving at Pearl Harbor in July. Shortly thereafter it earned the honor of being the first American battleship to go on the offensive against the Japanese.

The Showboat spent much of the war escorting carrier groups around the South Pacific, and was one of the only active American battleships in this theater in the early years of the war. She participated in numerous landing campaigns, from Guadalcanal in 1942 to Tokyo Bay in 1945. The USS North Carolina was added to the National Register of Historic Places in 1982 and named a National Historic Landmark in 1986.

The USS North Carolina is permanently located in Wilmington, North Carolina. Although looted for parts to repair other battleships in the 1980s, it is largely intact and much of it can be toured. As of the time of this writing the North Carolina was scheduled to undergo a major restoration, during which parts of the ship may be closed to visitors.

---

97. HANNAH BLOCK HISTORIC USO CENTER

120 South Second Street, Wilmington, North Carolina, 28401

Site Type: USO Center
Conflict: World War II
Dates: Opened in 1941
Designations: National Register of Historic Places
Web: www.wilmingtoncommunityarts.org (official website)

The Hannah Block Historic USO Building is one of the best preserved facilities of its kind. Built to host thousands of servicemen who passed through Wilmington during World War II, it now serves as a local community center. However, in honor of its historic past, the building maintains a gallery with art and artifacts that recall its history.

The United Service Organizations was founded in 1941 to provide assistance and entertainment to members of the armed forces. It was extremely active throughout World War II, with thousands of facilities around the country. There were several of these in the Wilmington area, which was surrounded by military bases and shipbuilding facilities. The most famous was the USO club at Second Street, where local celebrity and pianist Hannah Block entertained the troops.

The Hannah Block USO Center closed down after the war, and in 1948 was converted to use as a community center. A gallery in the center is now reserved for an exhibit which includes a variety of objects from newspapers to uniforms. The facility's original piano, which was played on many occasions by Hannah Block, is on display here as well.

98. MOORES CREEK NATIONAL BATTLEFIELD

40 Patriots Hall Drive, Currie, North Carolina, 28435

Site Type: Battlefield - Battle of Moores Creek Bridge
Conflict: American Revolution
Dates: Battle fought on February 27, 1776
Designations: National Register of Historic Places, National Battlefield
Web: www.nps.gov/mocr (official website)

The Battle of Moores Creek Bridge was one of the largest battles of the American Revolution in the South during the early years of the war. Fought between large bands of Rebel and Loyalist militias near Wilmington, this early American victory effectively kept the British out of North Carolina until a much larger invading force under Cornwallis arrived a few years later.

In the months after the Battle of Lexington and Concord, word spread up and down the colonies that war with Great Britain was now at hand. Militia units began mobilizing everywhere. In North Carolina Loyalists were called to arms to secure the city of Wilmington as a friendly port for the British in the South. Almost immediately Patriot militiamen began mobilizing to stop them.

Although initially smaller, the rebel force was better organized and led. The two sides clashed at Moores Creek Bridge, and thanks to several Loyalist blunders, the battle quickly became a rout. An overwhelming and decisive victory for the Americans, who had a total of two casualties, the battle resulted in almost the entire Loyalist force being killed or captured. The Moore's Creek Bridge Battlefield was listed on the National Register of Historic Places in 1966 and designated a National Battlefield in 1980.

Moore's Creek National Battlefield, expanded in the 1980s, covers most of the battle site. The original road that ran through the area in colonial times is still there, as is the bridge. Markers note where key actions of the battle took place.

99. FORT RALEIGH NATIONAL HISTORIC SITE

1401 National Park Drive, Manteo, North Carolina, 27954

Site Type: Earthwork Fort
Dates: Originally completed in 1587
Designations: National Register of Historic Places, National Historic Site
Web: www.nps.gov/fora (official website)

Fort Raleigh was the first English fort ever to be constructed in what is now the United States. Built to protect the colony at Roanoke Island in North Carolina, it was named for the colony's sponsor, Sir Walter Raleigh of Elizabethan fame. While the original fort is long gone, a recreation now stands close to what is believed to be the original location. Excavations are ongoing.

The first English settlement in America was established at Roanoke Island in 1585. A small earthwork fort was completed a year or so later. The colony famously did not last long, and sometime before resupply ships arrived in 1590 the settlement and its fort were completely abandoned. No trace of the settlers was ever found, giving rise to the legend of the Lost Colony.

Roanoke Island remained largely abandoned until the 1860s when it became home to freed slaves. Interest in the history of the island was revived in the early 20th century, when live performances of Paul Green's play *The Lost Colony* became an annual tradition. Fort Raleigh was named a National Historic Site in 1941 and listed on the National Register of Historic Places in 1996.

The Fort Raleigh National Historic Site is believed to be on or very near the original location of the colony. Just beyond the visitor center is a reconstruction of the original earthwork fort, along with some sections of wooden stockade construction. Nearby at the Waterside Theater visitors can still see performances of *The Lost Colony*.

100. GUILFORD COURTHOUSE NATIONAL MILITARY PARK

2332 New Garden Road, Greensboro, North Carolina, 27410

Site Type: Battlefield - Battle of Guilford Courthouse
Conflict: American Revolution
Dates: Battle fought on March 15, 1781
Designations: National Register of Historic Places, National Historic Landmark District
Web: www.nps.gov/guco (official website)

The Battle of Guilford Courthouse was the last major military engagement in the southern colonies during the American Revolution. Fought between a British army under Cornwallis and a vastly larger American force under Nathanael Greene, the battle was an unexpected British victory, but a very costly one. The massive casualties suffered by the British forced them to abandon their field of victory and ultimately withdraw north into Virginia.

Following the American victories at King's Mountain and Cowpens, the badly mauled British army in the Carolinas became desperate for a victory in order to maintain support from the local Loyalists. In mid-March Cornwallis received intelligence of a possible opportunity to crush the American army which was encamped at the town of Guilford Courthouse. Despite being outnumbered more than two-to-one, the British marched against Greene's army.

The American force, in addition to being larger, was entrenched in strong positions when the British arrived. Despite this the British were able to drive the Americans from the town. The rebel army largely got away in good order despite being abandoned by many of their militiamen. The British took heavy casualties, ending Cornwallis' ability to continue the campaign. Guilford Courthouse National Military Park was added to the National Register of Historic Places in 1966 and named a National Historic Landmark District in 2001.

Guilford Courthouse National Military Park includes only part of the battlefield site, with some of the area developed. While there are markers throughout the park noting sites associated with the battle, the accuracy of the locations is somewhat questionable. As of the time of this writing park authorities were working on a survey to replace the current markers with greater accuracy. There are monuments to Nathanael Greene and other leaders who participated in the engagement.

---

101. FORT SUMTER NATIONAL MONUMENT

340 Concord Street, Charleston, South Carolina, 29401 (mainland visitor center)

Site Type: Masonry Fort
Conflict: American Civil War
Date: Mostly completed by 1861(never finished)
Designations: National Register of Historic Places, National Monument
Web: www.nps.gov/fosu (official website)

Fort Sumter was erected in the wake of the War of 1812 as a defense for the city of Charleston against the threat of naval assault. Substantially completed in the 1830s, construction was still ongoing at the outbreak of the American Civil War. Fort Sumter is perhaps best known for being the place where the first shots of the Civil War were fired on April 12, 1861. Under the Confederacy it withstood several assaults before finally being abandoned in 1865. The fort remained in service throughout most of the 19th century.

In December of 1860 South Carolina became the first state to secede from the Union. One of its first acts was to secure its own territory, and in January of the next year South Carolina sent a demand to the federal government for the transfer of Fort Sumter to state control. This was refused, and forces from both sides began to converge on Charleston. When the first Union Navy ships arrived in the area, South Carolina moved swiftly, beginning an assault on April 12. Thirty hours later the garrison of the fort surrendered. Amazingly, none of the defenders were killed during the battle.

The Confederacy maintained control of Fort Sumter throughout most of the war. The garrison fought off a major Union assault in 1863. It wasn't until the arrival of Sherman's army in South Carolina that the defenders were finally forced to abandon the fort. After the war was over it was re-occupied by the Union army and remained in service until the 1870s. In 1876 it was converted to use as a lighthouse. It was briefly used by the military again during the Spanish-American War. Fort Sumter was designated a National Monument in 1948 and added to the National Register of Historic Places in 1966.

Fort Sumter stands on a small, rocky island in the center of Charleston Harbor, where it could threaten shipping while being secure from a land assault. A no-nonsense blocky affair, Fort Sumter is essentially a massive, multi-tiered battery of cannon surrounded by a wall. Largely destroyed during the Civil War, most of the modern fort actually dates from restorations in the late 19th century. A variety of artillery pieces adorn the casements. The Fort Sumter museum is actually located on the mainland and features exhibits on the history of the years leading up to the war. The fort is only accessible by ferry.

---

102. THE CITADEL

171 Moultrie Street, Charleston, South Carolina, 29409

Site Type: Military College
Dates: Opened in 1842
Web: www.citadel.edu (official website)

The Citadel is one of only six senior military colleges, along with the academies, in the United States. Its full name is The Citadel, The Military College of South Carolina. Graduates of The Citadel have served in every American conflict since the Mexican-American War. During World War II virtually every living graduate served in uniform.

The Military College of South Carolina began as a military academy in 1842. Cadets from the school are credited with firing the first shots of the American Civil War when they manned a cannon battery to stop a resupply ship from reaching Fort Sumter. The college moved to its current location in 1922. The Citadel's most notable alumnus is arguably William Westmoreland, who commanded American forces in Vietnam in the 1960s.

The Citadel has several sites of interest, including the Daniel Library, which houses the Citadel Museum. Monuments and artifacts on the campus include armored vehicles to small naval vessels to aircraft. The Seraph Monument commemorates Anglo-American cooperation in World War II, while the Howie Carillon honors Major Thomas Howie who led the attack on St. Lo during the Normandy invasion.

---

103. CSS HUNLEY & WARREN LASCH CONSERVATION CENTER

1250 Supply Street, North Charleston, South Carolina, 29405

Site Type: Submarine
Conflict: American Civil War
Dates: Commissioned in 1863
Designations: National Register of Historic Places
Web: https://hunley.org (official website)

The CSS Hunley, full name the CSS H.L. Hunley, was the first submarine in the world to successfully sink a surface military vessel in combat. Although the Hunley was also sunk in that historic naval engagement, she was rediscovered well over a century later and eventually recovered. The Hunley is now kept on display at the Warren Lasch Conservation Center in Charleston.

Submarines were among the many technical military innovations that were introduced during the Civil War. The first to be successfully completed, tested and used in combat was the CSS Hunley. The Hunley, constructed by the Confederacy as part of their effort to break the Union blockade, was launched in 1863. During testing she sank twice, but was recovered and repaired both times.

The Hunley's combat career lasted one day. On February 17, 1864 she attacked the sloop-of-war USS Housatonic. Ramming the Union ship with a torpedo, the Housatonic sank. However, the blast of the torpedo took out the Hunley as well. Despite the loss the battle is counted as an important milestone in the history of naval warfare. The Hunley was rediscovered in the 1970s and raised in 2000. It was listed on the National Register of Historic Places in 1978.

The CSS Hunley is now kept at the Warren Lasch Conservation Center. It is preserved in a huge salt-water tank which can be viewed through glass. A slightly larger scale replica of parts of the Hunley, which visitors can explore, is also on display at the center.

---

104. USS YORKTOWN & PATRIOT'S POINT

40 Patriot's Point Road, Mount Pleasant, South Carolina, 29464

Site Type: Aircraft Carrier
Conflicts: World War II, Vietnam War, Cold War
Dates: Commissioned in 1943
Designations: National Register of Historic Places, National Historic Landmark
Web: www.patriotspoint.org (official website)

The USS Yorktown (CV-10) is the second oldest aircraft carrier currently afloat, although it has the lowest numeric designation. The Yorktown was the second Essex class carrier put into service during World War II. It participated in many of the key naval campaigns in the Pacific, notably the Battle of the Philippine Sea, and was among the ships which hunted down the Japanese battleship Yamato. The Yorktown also saw service during the Vietnam War and worked with NASA as a recovery ship for the Apollo Space Program.

Construction on the Yorktown began less than one week before the Japanese attack on Pearl Harbor. Following the disastrous naval losses of December 7, as well as the loss of its predecessor at the Battle of Midway, the Yorktown was rushed to completion and entered active service in 1943. The carrier was primarily involved in fighting in the

Central Pacific, with notable engagements at Guam, the Philippine Sea and Okinawa. After World War II the Yorktown returned to the West Coast and was decommissioned for a few years.

Beginning in 1953 and throughout the Cold War the Yorktown was periodically brought back into active service. Aircraft from the Yorktown were occasionally involved in air operations over Vietnam. The carrier helped to recover the Apollo 8 spacecraft in 1968 and was a filming location for several movies in the 1970s and 1980s. The Yorktown received its final decommission in 1970, and in 1975 became permanently moored as a museum ship at Patriot's Point near Charleston. The USS Yorktown was named a National Historic Landmark in 1980 and added the National Register of Historic Places in 1982.

Patriot's Point is home to the USS Yorktown as well as several other museum ships and sites of military interest. There are also launches that depart from here that go to Fort Sumter in Charleston Harbor. The Yorktown is home to two dozen historic aircraft as well as a replica of the Apollo 8 capsule. Other ships at Patriot's Point include the destroyer USS Laffey and the submarine USS Clamagore.

---

105. KINGS MOUNTAIN NATIONAL MILITARY PARK

2625 Park Road, Blacksburg, South Carolina, 29702

Site Type: Battlefield - Battle of Kings Mountain
Conflict: American Revolution
Dates: Battle fought on October 7, 1780
Designations: National Register of Historic Places, National Military Park
Web: www.nps.gov/kimo (official website)

The Battle of King's Mountain marked the turning point of the American Revolution in the South and, along with the subsequent Battle of Cowpens, the beginning of the end for the British. After a string of defeats, a disheartened band of American militia was confronted by a superior force of Loyalist militia. The Americans managed to completely outmaneuver their opponents, and after an hour-long battle the entire enemy force of over a thousand men was killed, wounded or captured. King's Mountain was the largest engagement of the war to be fought primarily by militia on both sides.

In 1778 the British conquest of Savannah kicked off the Southern phase of the American Revolution. In the two years that followed the British under Cornwallis won a string of victories that nearly brought all of the southern colonies back into Loyalist hands. By September 1780 the British had advanced deep into the Carolinas. Both sides raised significant militia forces for the campaign.

On October 7 a large force of Loyalist militiamen, separated from the British army, was surrounded by the rebels. Although outnumbered, the Americans caught the Loyalists almost completely by surprise. The result was a disaster for the British, with every one killed, wounded or captured. This paved the way to an even more decisive victory at Cowpens a few months later. Kings Mountain National Military Park was established in 1931 and listed on the National Register of Historic Places in 1966.

King's Mountain National Military Park is the largest Revolutionary War battlefield site in the South. It is part of the Overmountain Victory National Historic Trail, which marks the route from Tennessee to South Carolina along which the Patriot militia gathered for the battle. The main site of the engagement is marked by a commemorative obelisk. Patrick Ferguson, who commanded the Loyalist militia, is buried in the park.

106. COWPENS NATIONAL BATTLEFIELD

4001 Chesnee Highway, Gaffney, South Carolina, 29341

Site Type: Battlefield - Battle of Cowpens
Conflict: American Revolution
Dates: Battle fought on January 17, 1781
Designations: National Register of Historic Places, National Battlefield
Web: www.nps.gov/cowp (official website)

The Battle of Cowpens was the last major British attempt at an offensive campaign in the South before their final withdrawal from the region. Following their unexpected defeat at Kings Mountain a few months earlier, the British regrouped and attempted to hunt down the troublesome American army in South Carolina. What followed was a total disaster for the British and the rout of one of its largest armies.

After Kings Mountain the British commander Cornwallis decided to make one last all-out attempt to crush the Americans in the Carolinas once and for all. He dispatched a part of his force under the infamous Banastre Tarleton to track down the American army and maneuver them into a trap between the British armies. Instead the Americans retreated to Cowpens where, suddenly reinforced by a large body of Patriot militia, they set a trap for Tarleton's force.

The two armies met at Cowpens on January 17. In one of the most famous tactics in American military history, the American commander Daniel Morgan set up the unreliable militia in the front, had them fire two volleys at the British and then pretend to flee. As the British pursued them, they suddenly found themselves face to face with previously concealed American Continentals. The charge became a disorganized retreat. By the time the fighting was done, the entire British force was dead, wounded or captured. This defeat was the final nail in the coffin for the British campaign in the South. The Cowpens National Battlefield was established in 1929 and was listed on the National Register of Historic Places in 1966.

Cowpens National Battlefield, a former cow pasture, incorporates the entire site of the battle. Not as large as nearby King's Mountain, the entire site can be walked in a relatively short time. A large stone pillar in front of the visitor's center commemorates those who fought in the battle. Another monument in the middle of the battlefield erected by the Washington Light Infantry of Charleston marks the location where the fighting ended.

---

107. FORT PULASKI NATIONAL MONUMENT

101 Fort Pulaski Road, Savannah, Georgia, 31410

Site Type: Masonry Fort
Conflict: American Civil War
Dates: Originally completed in 1847
Designations: National Register of Historic Places, National Monument
Web: www.nps.gov/fopu (official website)

Fort Pulaski was one of the major fortifications constructed after the War of 1812 designed to protect America's coastal cities. One of the largest, Fort Pulaski guarded the approaches to the port of Savannah. It was completed in 1847, just a few years ahead of the Civil War. The fort remained in service until the early 20th century.

Fort Pulaski was one of the largest of the coastal forts to be completed prior to the Civil War. One of the officers involved in overseeing its construction was Robert E. Lee. Essentially garrisoned by only a few caretakers after its completion, Fort Pulaski was seized by Confederate forces when Georgia seceded from the Union.

In April 1862 the fort was retaken by Union troops after a brief siege with extremely powerful new artillery. With the fall of Fort Pulaski, Savannah was effectively lost as a port of commerce for the South. The Confederacy never recaptured it. Fort Pulaski was named a National Monument in 1924 and listed on the National Register of Historic Places in 1966.

Fort Pulaski is located on Cockspur Island at the mouth of the

Savannah River. Repaired during the war and maintained for a long period afterwards, it is in generally better condition than most of America's other surviving coastal forts. However, some of the outer wall still exhibits damage from the 1862 artillery siege. The moat is one of the only such structures still preserved in the United States.

---

108. NATIONAL MUSEUM OF THE MIGHTY EIGHTH AIR FORCE

175 Bourne Avenue, Pooler, Georgia, 31322

Site Type: Museum
Dates: Opened in 1996
Web: www.mightyeighth.org (official website)

The National Museum of the Mighty Eighth Air Force is a military aviation museum specifically dedicated to the history and planes of the United States Eighth Air Force, and in particular their participation in World War II. While not as large as some of the other national military aviation museums, it does offer a more in depth look at one specific command and its history.

The Eighth Air Force was formed early in 1942, shortly after America's entry into World War II. Organized for operations against Nazi Germany, the Eighth was the anchorman of the American strategic bombing campaign against enemy military and industrial targets. Its fighters and bombers played a role in almost every battle in the European theater and were instrumental in Germany's surrender. The Eighth Air Force has been active in almost every American conflict for the last seven decades.

The National Museum of the United States Air Force has both indoor and outdoor exhibits. Indoor exhibits include galleries that tell the history of the Eighth Air Force, its predecessors and those who have served. Outside there are planes on display from both World War II and the post-war era. At the time of this writing a B-17 Flying Fortress was being repaired on full display in an indoor wing of the museum.

109. NATIONAL INFANTRY MUSEUM AND SOLDIER CENTER & FORT BENNING

1775 Legacy Way, Columbus, Georgia, 31903

Site Type: Military Base, Museum
Dates: Base established in 1909
Web: www.nationalinfantrymuseum.org (official website)

The National Infantry Museum and Soldier Center is devoted to telling the story of the American soldier, from the earliest days of the Republic to the 21st century. Originally housed at nearby Fort Benning, it is home to an expansive collection of military artifacts from virtually all of America's wars. There are also several memorials and monuments on site, as well as exhibits related to the history of Fort Benning.

Fort Benning is one of the largest and most important military installations in the United States. Founded in 1909, it is currently home to the United States Army Infantry School, Armor School and Airborne School. A number of Army units are based here, including infantry, armor and rangers. For many years this historic army base was also home to the National Infantry Museum. The museum was moved to a dedicated and much larger facility in 2009.

The National Infantry Museum is actually next door to Fort Benning. It hosts a series of exhibits on American Infantry through the decades, with a focus on the 20th century. There is also an exhibit on the related history of cavalry and armor. Also on site is a recreation of 1940s Fort Benning, with barracks and other buildings as they appeared during World War II. Monuments at the museum honor those who fought in Vietnam and the War on Terror.

110. NATIONAL CIVIL WAR NAVAL MUSEUM

1002 Victory Drive, Columbus, Georgia, 31901

Site Type: Museum
Conflict: American Civil War
Dates: Opened in 1962
Web: www.portcolumbus.org (official website)

The National Civil War Naval Museum is the only museum in the United States devoted entirely to naval combat during the Civil War. It documents, among other things, the beginning of the ironclad age, which ushered the American Navy into the modern era. The museum is home to one of the largest collections of recovered and recreated warships of the conflict.

Founded in 1962 and moved to its current location in 2001, the Civil War Naval Museum thoroughly traces the history of the war at sea from its outset to its ending. The museum houses the largest collection of Civil War naval artifacts in the United States, and includes everything from naval artillery to signal flags.

The National Civil War Naval Museum has a number of ships on display both indoors and outdoors. The most important original ship is the ironclad CSS Jackson, of which most of the hull was recovered and is now on display here. The USS Hartford and a remnant of the CSS Chattahoochee are also on display. There is a partial recreation of the ironclad USS Monitor as well. Outside the museum is a full recreation of the USS Water Witch, a Union vessel that was later seized and used by the Confederacy.

111. ANDERSONVILLE NATIONAL HISTORIC SITE

496 Cemetery Road, Andersonville, Georgia, 31711

Site Type: POW Camp
Conflict: American Civil War
Dates: Opened in 1864
Designations: National Register of Historic Places, National Historic District, National Historic Site
Web: www.nps.gov/ande (official website)

Andersonville Prison, formerly known as Camp Sumter, is the best known and most infamous prisoner of war camp in American history. Constructed in 1864, it became the chief prisoner of war camp in the South for the last year of the conflict. During its brief existence more than forty thousand Union soldiers were incarcerated here, of which more than one in four died from the horrific living conditions.

By early 1864, as victory for the North was looking increasingly inevitable, the South constructed a massive new POW camp at Andersonville. Many Union prisoners were relocated here from other camps as federal troops began advancing on all fronts. The camp was completed in April of 1864, and within two months nearly doubled in size. Andersonville became horribly overcrowded, resulting in widespread suffering, rampant disease and starvation.

Approximately forty-five thousand Union prisoners were kept here from April 1864 until its liberation in May 1865. During that time span about thirteen thousand men died from privation and exposure. Although the South did make an effort to arrange a prisoner exchange in order to alleviate the overcrowding, this was rejected by the North as it would have extended the war. The offer was rescinded once Sherman began his March to the Sea. After the war the camp commandant was executed for war crimes. Andersonville was listed on the National Register of Historic Places and named a National Historic Site in 1970.

Andersonville National Historic Site is located at on the grounds of the original camp, but much of it is recreated. Sections of log stockade walls, watchtowers, tents and other makeshift shelters offer a glimpse of the living conditions in 1864. Many of the prisoners who died here are buried at the Andersonville National Cemetery. In 1998 Andersonville also became home to the National Prisoner of War Museum, which honors all those from every conflict who were kept as POWs by enemy states.

---

112. KENNESAW MOUNTAIN NATIONAL BATTLEFIELD PARK

900 Kennesaw Mountain Drive, Kennesaw, Georgia, 30152

Site Type: Battlefield – Battle of Kennesaw Mountain
Conflict: American Civil War
Dates: Battle fought on June 18-July 2, 1864
Designations: National Register of Historic Places, National Battlefield Park
Web: www.nps.gov/kemo (official website)

The Battle of Kennesaw Mountain was one of the largest and most complex engagements fought during the Union campaign to capture the city of Atlanta. Although a victory of sorts for the Confederacy, it did little to stop the Northern juggernaut, which in the end simply went around the Southern defensive positions, forcing the rebels to fall back anyway.

By the summer of 1864 a large Union army under the command of William T. Sherman was rapidly advancing from Tennessee towards Atlanta. Defending the approach to the city was Joe Johnson with about half as many men. Johnson prepared his major defense at Kennesaw Mountain where the two armies met towards the end of June. The engagement was characterized by nearly two weeks of maneuvering and skirmishing.

The heaviest fighting took place on June 27, when the Union army made its greatest effort to dislodge the defenders entrenched around the city of Marietta, with no success. Casualties were moderate considering the size of the battle. After Kennesaw Mountain Sherman resumed his previous strategy of simply going around the Confederate army, and by early September Atlanta had fallen. Kennesaw Mountain was named a National Battlefield Park in 1935 and was listed on the National Register of Historic Places in 1966.

Kennesaw Mountain National Battlefield Park is one of the largest preserved battle sites from the Civil War. The battle took place over a huge area, and only part of the field is incorporated into the park. There are several separate areas in the park, with the main visitor center located at the northern end. Trail markers throughout the park note points of interest related to the battle.

---

113. CHICKAMAUGA AND CHATTANOOGA NATIONAL MILITARY PARK

3370 LaFayette Road, Fort Oglethorpe, Georgia, 30742

Site Type: Battlefield - Battles of Chickamauga and Chattanooga
Conflict: American Civil War
Dates: Battles fought on September 18-20, 1863 & September 21-November 25, 1863
Designations: National Register of Historic Places, National Military Park
Web: www.nps.gov/chch (official website)

The Battle of Chickamauga and the subsequent Chattanooga campaign was the bloodiest engagement of the civil war south of Virginia. Chickamauga was a short-lived victory for the South which briefly impeded the advance of the Union army into Georgia. However after receiving reinforcements the North immediately resumed the offensive in a series of engagements in the Chattanooga campaign that lasted for several months. By the end of this series of battles the Confederates were defeated, Chattanooga secured by the Union and the way opened for a march on Atlanta.

Throughout 1863, while Union and Confederate forces in Northern Virginia continued their stalemate around Washington, the United States Army of the Cumberland made significant gains in Tennessee. In September 1863, after capturing Chattanooga, they crossed into Georgia where they were met by a Confederate army recently reinforced by units from Virginia. After three days of vicious fighting at the Battle of Chickamauga, where both sides suffered enormous casualties, the Union Army was forced to withdraw to Chattanooga.

For the next month the Confederate Army laid siege to Chattanooga. But in late October a large relief force led by Grant and Sherman broke the siege and immediately began pressing the Confederates. Finally between November 23 and 25 the Union army retook the surrounding area and secured Lookout Mountain. These actions drove the Confederate army from Tennessee and laid the groundwork for Sherman's invasion of Georgia. Chickamauga and Chattanooga National Military Park was established in 1890. It was listed on the National Register of Historic Places in 1966.

Chickamauga and Chattanooga National Military Park is the oldest battlefield park in the National Park system and was the first to be opened. It consists primarily of the Chickamauga Battlefield along with a few other points of interest of the Chattanooga campaign, including Moccasin Bend and Missionary Ridge. Lookout Mountain is home to several other related sites and is the most visited part of the park.

DEEP SOUTH

114. CASTILLO DE SAN MARCOS NATIONAL MONUMENT

1 South Castillo Drive, St. Augustine, Florida, 32084

Site Type: Masonry Star Fort
Dates: Originally completed in 1695
Designations: National Register of Historic Places, National Monument
Web: www.nps.gov/casa (official website)

The Castillo de San Marcos is the oldest stone fortress still standing in North America. Constructed by the Spanish empire to guard the northern frontier of its New World territories, the fort became a focal point in the colonial wars between Spain and England. It endured several sieges in the early 18th century before being ceded to England in 1763. The Castillo de San Marcos changed hands several more times between England and Spain before finally being turned over to the United States in 1819. The Castillo was used throughout the 19th century as a prison before being decommissioned in 1900.

Seven decades after Christopher Columbus discovered the New World, the Spanish founded the first permanent European settlement in what would become the United States at St. Augustine. Shortly after its founding, the Spanish erected a wooden fort to protect the colony. During the next two centuries Florida became a focal point in the colonial wars between the Spanish, English and French. St. Augustine was sacked several times, notably by Sir Francis Drake in 1586 and by pirates in 1668. After the latter attack, the Spanish monarchy decided to build an enormous stone fort to protect the colony.

Finished a few years later, the Castello de San Marco soon gained a reputation as one of the most formidable fortresses in the Americas. It withstood English sieges in 1702 and 1739, and it wasn't until 1763 that England took control of St. Augustine following the Treaty of

Paris. After the American Revolution, Britain ceded Florida back to Spain, who in turn ceded it to the United States in 1821. During the Seminole Wars the fort was used as a prison for captive Native Americans. The Castillo de San Marco was declared a National Monument in 1924 and added to the National Register of Historic Places in 1966.

The Castillo de San Marcos is a nearly perfect example of a 17th century star fort. Overlooking the convergence of the Matanzas River and the St. Augustine Inlet, it guards the waterway approaches to the city. The thick walls built from coquina stone and the diamond shaped towers in the corners are in nearly pristine condition and boast a battery of antique cannon peeking out from the parapets. Most of the Castillo's interior rooms, including the barracks and a well secluded powder magazine, have been restored and are open to visitors.

---

115. BAY OF PIGS MUSEUM

1821 Southwest 9th Street, Miami, Florida, 33135
(relocating at the time of this writing)

Site Type: Museum
Conflict: Cold War
Date: Opened in 1988
Web: None Available

The Bay of Pigs Museum, official name the Brigade 2506 Museum and Library, is a small private museum currently located in the Little Havana neighborhood of Miami. Many of those involved in the famous, ill-fated attack resided or currently reside in this area. The museum features a collection of artifacts from the 1961 invasion and exhibits on those who fought to liberate their homeland from Communist rule.

In the late 1950s revolutionaries under the leadership of Fidel Castro overthrew the government of Cuba. Having a Communist government in power less than a hundred miles from Florida was deemed to be an existential threat to the United States, and Washington began making plans to overthrow Castro. Under the direction of the CIA, anti-Castro Cubans living in Miami were recruited and trained for an invasion of the island.

The subsequent campaign, which began on April 17, 1961 and ended a few days later, was an unmitigated disaster. Despite substantial support from the United States armed forces the invasion quickly collapsed. Over a hundred members of Brigade 2506 were killed and almost all of the rest were taken prisoner. The invasion led to a worsening of relations between the United States and Cuba, and was a contributing factor in the Cuban Missile Crisis in 1962.

The Bay of Pigs Museum houses a large collection of artifacts related to the event. For many years members of the museum staff who fought in the invasion were available to give personal accounts, though these are now in their eighties. Not too far from the museum is the Bay of Pigs Monument honoring those who died in the campaign. As of the time of this writing the Bay of Pigs Museum is planned to be moved to a new location in Hialeah Gardens.

---

116. OCALA/MARION COUNTY VETERANS MEMORIAL PARK

2601 East Fort King Street, Ocala, Florida, 34470

Site Type: Monuments – Numerous
Dates: Opened in 1997
Web: www.marioncountyfl.org/departments-agencies/departments-o-z/veterans-services/ocala-marion-county-veterans-memorial-park (official website)

The Ocala/Marion County Veterans Memorial Park is the largest and most diverse military memorial site not run by the federal government. Located near the center of Ocala, this pleasant public park is home to hundreds of monuments, markers, dedicated benches and thousands of bricks, all of which commemorate American veterans or some event of American military history. Opened in 1997, the park was rededicated in 2005 and continues to see new monuments added periodically.

The park began as a project of local veterans in the 1980s. It was conceived both as a place to honor veterans and also to commemorate many conflicts in American history and even earlier. There are monuments that identify events as far back as four centuries ago, to the first conflicts between European settlers and local tribes. The park is staffed entirely by volunteers, most of whom are veterans, and there are veteran's service facilities on site.

The Veterans Memorial Park is expansive, with many sections, including a Medal of Honor plaza and a memorial pavilion. The thousands of bricks throughout the park are mostly dedicated to individual veterans. At the time of this writing there were nearly two hundred monuments, plaques, benches, walls and flagpoles commemorating everything imaginable including wars, battles, service branches, military units, leaders, heroes, warships, forts, civilian institutions and more.

---

117. NATIONAL NAVAL AVIATION MUSEUM

1878 South Blue Angel Parkway, Pensacola, Florida, 32507

Site Type: Museum
Dates: Opened in 1962
Web: www.navalaviationmuseum.org (official website)

The National Naval Aviation Museum is the one of the finest military air museums in America. Located at the Naval Air Station in Pensacola, which is also home to the world famous Blue Angels fighter squadron, it is the largest museum dedicated exclusively to naval air power in the United States. It is also home to the Naval Aviation Hall of Honor, which commemorates such luminaries as Neil Armstrong and George Bush Sr. to leaders like Admiral William Halsey and inventors like Igor Sikorsky.

The Naval Air Station at Pensacola, founded in 1913, is the oldest such military base in America. The base was one of the navy's most important training centers during both World Wars. It has been home to America's most famous fighter squadron, the Blue Angels, since the 1940s. The museum was opened on the site of the base in 1962 and was moved to its current location in 1974.

The National Naval Aviation Museum has a collection of over one hundred and fifty aircraft, primarily featuring planes and artifacts from World War II and the Cold War. Virtually every model of naval aircraft ever used in the United States is represented, with some odds and ends thrown in, including one of the last surviving World War I German Fokker DVIIs and British Sopwith Camels. Also on display is a helicopter which formerly served as Marine One. A quartet of Blue Angel jets is on display suspended from the main atrium. On certain days lucky visitors may actually get to see a Blue Angels training flight.

---

118. SAN JUAN NATIONAL HISTORIC SITE

501 Calle Norzagaray, San Juan, Puerto Rico, 00901

Site Type: City Fortifications
Dates: Originally completed in 16th Century
Designations: UNESCO World Heritage Site,
National Register of Historic Places
Web: www.nps.gov/saju (official website)

The San Juan National Historic Site is an immense historic district that incorporates most of the surviving colonial-era defenses of the city of San Juan on the island of Puerto Rico. San Juan, one of the most heavily defended cities in the New World in the 16th, 17th and 18th centuries, boasts more surviving fortifications than any other city in the Americas. The San Juan National Historic Site includes the Castillo San Felipe del Morro, the Castillo San Cristobal and the Fortin San Juan de la Cruz, as well as the remaining city walls.

During the colonial period Spain was the most prolific builder of massive stone fortresses in the Caribbean. As the first major port Spanish ships would reach after crossing the Atlantic Ocean, San Juan was a critical transportation center and required serious defenses. By the end of the 16th century the city and port were protected by thick walls and daunting fortresses, the largest and most imposing of which was the Castillo San Felipe del Morro.

Between the late 16th century and 1800, San Juan and its forts endured attacks by the English and the Dutch, as well as pirates, including one attack in 1595 by Francis Drake. San Juan remained one of the last surviving Spanish bastions in the Caribbean right up until 1898 when the island was ceded to the United States. The city's fortifications were partially upgraded for use by the United States military in the 20th century. The San Juan National Historic Site was listed on the National Register of Historic Places in 1966, and named a UNESCO World Heritage Site in 1983.

Fort San Felipe Del Morro is an immense, multi-level citadel built on a strategic outcropping of rock that guards the entrance to San Juan Bay. The top level of the citadel backs up against a plateau on which stands the city of San Juan. Of all of the colonial fortresses on the island, Fort Del Morro most resembles a late medieval European castle, and looks as if it had been built by crusaders rather than conquistadors. It now houses a museum. The smaller Fortin San Juan de la Cruz stands opposite Del Morro on Palo Seco Island, while the Castillo San Cristobal guards the landward approach to the city.

---

119. USS ALABAMA & BATTLESHIP MEMORIAL PARK

2703 Battleship Parkway, Mobile, Alabama, 36602

Site Type: Battleship
Conflict: World War II
Dates: Commissioned in 1942
Designations: National Register of Historic Places, National Historic Landmark, Alabama Register of Landmarks and Heritage
Web: www.ussalabama.com (official website)

The USS Alabama (BB-60), or *Lucky A*, was the first new American battleship to be launched after the Japanese attack on Pearl Harbor. Commissioned in 1942, the Alabama's first assignment in the war was to guard supply convoys across the Atlantic. The rest of its service was in the Pacific theater where it generally served as an escort ship to various aircraft carriers. It participated in many of the major engagements towards the end of the war, notably at the Battle of the Philippine Sea. The Alabama received the Lucky moniker due to the fact that it suffered virtually no damage and almost no casualties during the war.

The Alabama spent much of its first year on active duty in the North Atlantic. Working with the British fleet, it was involved in escort duty for convoys, and was used to secure the northern sea route around Scandinavia to the Russian port of Murmansk. By the end of 1943 the Alabama had transferred to the Pacific Theater.

For nearly two years the Alabama participated in numerous campaigns in the South Pacific including the Marianas, the Caroline Islands and the Philippine Sea. She survived a typhoon in December of 1944 that sank several other American ships. The Alabama arrived in Tokyo Bay just a few days after the Japanese surrender. Returning to the United States she was decommissioned in 1947 and converted into a museum ship in 1964. The USS Alabama was designated a National Historic Landmark in 1984.

Battleship Memorial Park in Mobile Bay is home to the USS Alabama as well as the submarine USS Drum. Of all of the World War II battleships converted into museums, the Alabama is in perhaps the best shape due to the fact that it suffered very little damage during the war and was permanently decommissioned so quickly afterwards. The Alabama has a very minor list caused by Hurricane Katrina in 2008.

120. TUSKEGEE AIRMEN NATIONAL HISTORIC SITE

1616 Chappie James Avenue, Tuskegee, Alabama, 36083

Site Type: Airfield
Conflict: World War II
Dates: Completed in 1940
Designations: National Register of Historic Places, National Historic District
Web: www.nps.gov/tuai (official website)

The Tuskegee Airmen were among the most storied military aviators of World War II. Based in Tuskegee, Alabama, this squadron of African-American pilots overcame an historic barrier that helped to pave the way for the desegregation of the armed forces in the postwar era. The field and facilities where the first squadron was formed and trained is now home to a museum honoring their achievements.

Although the army air corps had existed since World War I, African Americans were barred from participating in military aviation until 1941. In that year a number of African-Americans began to train in various capacities at Moton Field in what became known as the Tuskegee Experiment. Pilots, ground crew and other support staff were all trained here, attracting some of the finest and bravest men from around the country.

Hundreds of African American pilots trained at Tuskegee went on to become distinguished combat fliers in World War II. The base was closed and the airstrip turned over to civilian control in 1946. Tuskegee Airmen National Historic Site was named a National Historic District and listed on the National Register of Historic Places in 1998.

Tuskegee Airman National Historic Site is located at Moton Field, which still operates as a civilian airport. The museum is located in Hangar One where the original Tuskegee Airmen training took place. Several training and combat planes used by the squadron are on display here, as well as artifacts and photographs from the time that Moton was an active military airfield.

121. HORSESHOE BEND NATIONAL MILITARY PARK

11288 Horseshoe Bend Road, Daviston, Alabama, 36256

Site Type: Battlefield – Battle of Horseshoe Bend
Conflicts: War of 1812, Early American Indian Wars
Dates: Battle fought on March 27, 1814
Designations: National Register of Historic Places, National Military Park
Web: www.nps.gov/hobe (official website)

The Battle of Horseshoe Bend was the decisive military engagement of the Creek War, which in turn was part of the greater conflict between the United States and Great Britain in the War of 1812. It was also one of the most significant Native American attempts to halt settler expansion in the South. While it was a victory over the British, it also represented another sad chapter for the Native Americans desperately fighting to hold on to their territory.

During the War of 1812 many tribes living on the American frontier, largely in the Louisiana Territory, saw the British as an ally that could be used to help the stop the inexorable and often violent march of American settlers into the west. In the South large numbers of Creek warriors known as the Red Sticks were galvanized into action and fought with the goal of preventing further encroachments. Early in 1814 future president Andrew Jackson led an army into Alabama to suppress the uprising.

The two armies met on March 27 at Horseshoe Bend, where the Red Sticks had entrenched themselves in a strong position on a hill. However, they were significantly outnumbered, and despite all their efforts, the position was overrun by Jackson's army. The result was a massacre. No quarter was given, and almost every member of the Red Sticks was killed or wounded. At the time it was hailed as a great victory for Jackson. Today this is seen in a different light as part of the tragedy of the Native American people. Horseshoe Bend National

Military Park was established in 1956 and added to the National Register of Historic Places in 1813.

Horseshoe Bend National Military Park serves a dual purpose: to remember this battle, and to commemorate the nearly one thousand Creek Indians who died here trying to protect their territory. Most of the battlefield site is preserved and there are markers noting where events of the engagement took place. There are monuments and memorials on site commemorating both sides.

---

122. AFRICAN AMERICAN MILITARY HISTORY MUSEUM

305 East 6th Street, Hattiesburg, Mississippi, 39401

Site Type: Museum
Dates: Opened in 2009
Designations: National Register of Historic Places, Mississippi State Landmark
Web: www.hattiesburguso.com (official website)

The African American Military History Museum in Hattiesburg is one of the only institutions of its kind in the United States. Built in a former USO facility, this museum has exhibits on the entire history of the African American military experience, from the Revolution to the War on Terror. Badly damaged by a tornado in 2013, the museum has recently reopened with beautifully restored exhibits that tell an important story.

The museum itself is in an historic building. Built in 1942, it housed a USO club for use by African-American soldiers. Relatively few such clubs were built for non-whites, and fewer have survived to the present day. There are exhibits on the USO in the lobby, and former rooms of the USO club now host nearly a dozen galleries of exhibits.

The African American Military History Museum has several excellent exhibits. Probably the most important and interesting are those on the Buffalo Soldiers who served on the American frontier,

African-American soldiers in World War II, and the Desegregation of the military in the 1940s. There is also a Hall of Honor which commemorates African American soldiers from the local area that have served and sacrificed for their country.

---

123. BRICE'S CROSSROADS NATIONAL BATTLEFIELD SITE

260 Bethany Road, Guntown, Mississippi, 38849

Site Type: Battlefield - Battle of Brice's Crossroads
Conflict: American Civil War
Dates: Battle fought on June 10, 1864
Designations: National Register of Historic Places, National Battlefield Site
Web: www.nps.gov/brcr (official website)

The Battle of Brice's Crossroads was the first of two battles that took place in the vicinity of Tupelo, Mississippi in the summer of 1864. It was fought between a Union force trying to protect the federal supply lines in Tennessee supporting Sherman's Atlanta campaign, and a Confederate force under Nathan Bedford Forrest trying to disrupt the supply lines. A victory for the South, Brice's Crossroads managed to delay Union operations in Tennessee for more than a month.

In May of 1864 the main Union force under Sherman began to move south from Tennessee towards Atlanta. One of the most important campaigns of the war, Sherman's offensive was threatened by the long and highly exposed supply lines that went all the way back to Nashville. In fact one of his biggest concerns was an all-cavalry Confederate force in Northern Mississippi which could easily be moved against him.

Sherman dispatched an army to Northern Mississippi to prevent this from occurring. The two sides met at Brice's Cross Roads. Despite significantly superior numbers, the Union force was routed with heavy casualties. This left the Confederate cavalry as a major threat for the next month. Brice's Cross Roads became a National Battlefield Site in 1933 and listed on the National Register of Historic Places in 1966.

Brice's Cross Roads National Battlefield Site is a relatively small park, and only encompasses a fraction of the original battlefield. The main site of interest is the impressive monument which commemorates the fallen of both sides, which is located near the visitor's center.

---

124. TUPELO NATIONAL BATTLEFIELD

2005 Main Street, Tupelo, Mississippi, 38801

Site Type: Battlefield - Battle of Tupelo
Conflict: American Civil War
Dates: Battle fought on July 14-15, 1864
Designations: National Register of Historic Places, National Battlefield Site
Web: www.nps.gov/tupe (official website)

The Battle of Tupelo was the second major military engagement in the vicinity of Tupelo, Mississippi within a span of five weeks. This time a victory for the North, Tupelo was critical in countering a Confederate cavalry attack into Tennessee that threatened to disrupt Union supplies to Sherman's Atlanta campaign in Georgia.

Following their defeat at Brice's Crossroads in June, the Union army redoubled its efforts to eliminate the Confederate threat in Northern Mississippi. Fortunately for the North the Confederate forces did not press into Tennessee during the interim. Within a month the two armies in the area, both having doubled in size, came to blows again near Tupelo.

This time the Union army was able to outmaneuver the Confederates and take up a stronger position on the battlefield. The result was heavy losses for the South, forcing them to retreat from the field. Although the Confederate army remained intact, their devastated ranks no longer posed a serious threat the Union army in Tennessee. The Tupelo National Battlefield was established in 1929 and listed on the National Register of Historic Plces in 1966.

The Tupelo National Battlefield encompasses a part of the battle site. The main point of interest is the imposing Battle of Tupelo Memorial, a large monument commemorating those who fought on both sides of the engagement. It is flanked by a pair of cannon. Markers note other key places around the battlefield.

---

125. VICKSBURG NATIONAL MILITARY PARK

3201 Clay Street, Vicksburg, Mississippi, 39183

Site Type: Battlefield - Siege of Vicksburg
Conflict: American Civil War
Dates: Siege fought from May 18-July 4, 1863
Designations: National Register of Historic Places, National Military Park
Web: www.nps.gov/vick (official website)

The Siege of Vicksburg was the culmination of the Union effort to control the Mississippi River and effectively knock the western states out of the war. Along with the defeat of Lee at Gettysburg, which took place on the same day as the end of the siege of Vicksburg, the surrender here marked the beginning of the end for the Confederacy. In terms of enemy casualties inflicted, it was also one of the North's most overwhelming victories of the war.

Control of the Mississippi River was one of the key strategic goals of the Union from the outset of the conflict. Fighting along the river began as early as 1861, and the vital port city of New Orleans was seized in the summer of 1862. Within a year most of the river was in Northern hands. The last holdout was the well defended city of Vicksburg. Confederate artillery set on high bluffs here completely controlled river traffic while remaining invulnerable to naval fire.

Several attempts were made to take Vicksburg, both from the land and the river. In late May of 1863 Ulysses S. Grant, commander of the Union army, changed tactics and surrounded Vicksburg. On July 4, after seven weeks of siege and no sign of relief in sight, supplies in

Vicksburg ran out and the city surrendered. Without control of the Mississippi river, the western states of Texas, Louisiana and Arkansas were cut off from the rest of the Confederacy and effectively removed from the war. Vicksburg National Military Park was created in 1899 and listed on the National Register of Historic Places in 1966.

Vicksburg National Military Park is huge, encompassing numerous sites related to the siege, including some across the Mississippi River in Louisiana. Extensive sections of earthworks adorned with cannons crisscross the park. Vicksburg has among the largest number of monuments of any battlefield site in the United States, the most notable of which is the Illinois State Memorial. Also here is the recovered and partially restored USS Cairo, one of the only surviving Union ironclad ships.

---

126. MANHATTAN PROJECT NATIONAL HISTORICAL PARK

300 South Tulane Avenue, Oak Ridge, Tennessee, 37830

Site Type: Military Research Facility
Conflict: World War II
Date: Opened in 1942
Designations: National Historical Park
Web: www.nps.gov/mapr (official website)

The Manhattan Project was one of the greatest and most terrifying scientific endeavors ever undertaken by mankind. Begun in 1942, this top secret research into atomic energy led to the creation of the world's first nuclear weapons. This development put a quick end to World War II and ushered in both the Atomic Age as well as the Cold War. Over a dozen locations were involved in the Manhattan Project, with the main sites being at Oak Ridge, Tennessee; Los Alamos, New Mexico; and Richland, Washington.

At the outbreak of World War II some of the world's leading scientists approached President Roosevelt with the possibility of developing atomic weapons and dire warnings that Nazi Germany was already attempting to do so. Roosevelt authorized the Manhattan Project in October 1941, just a few months before America entered the war. By 1942 the project was well underway, with the main work going on at Oak Ridge.

Less than three years later, on July 17, 1945, the first atomic weapon was tested at Los Alamos in New Mexico. Less than a month after that, atomic weapons were dropped on the Japanese cities of Hiroshima and Nagasaki. Japan surrendered shortly thereafter. The Manhattan Project was a military success, albeit a foreboding one, and was formally disbanded in 1947. The Manhattan Project National Historic Park was established in 2015.

The Manhattan Project National Historic Park consists of three separate facilities at Oak Ridge, Los Alamos and Richland. The most interesting location of the three is the Oak Ridge site. The Los Alamos site where the first bombs were detonated was not publically accessible at the time of this writing. Associated with the American Museum of Science and Energy, bus tours of the Oak Ridge site are available from the museum and include one of the original reactors and the electromagnetic separation plant.

---

127. SERGEANT ALVIN C. YORK STATE HISTORIC PARK

2609 North York Highway, Pall Mall, Tennessee, 38577

Site Type: Museum
Conflict: World War I
Dates: Opened in 1967
Designations: National Register of Historic Places, National Historic Landmark District, Tennessee State Park
Web: http://tnstateparks.com/parks/sgt-alvin-c-york (official website)

Alvin York is one of the most famous soldiers in American history. Raised a pacifist, York was drafted into the army during World War I and went on to become one of the most decorated combat soldiers of all time. In addition to the Congressional Medal of Honor, York received some of the highest decorations possible from several other countries. York's life was later immortalized by Hollywood, helping to make him a household name.

Alvin York entered into service in the United States Army in 1917 despite being a conscientious objector. The action for which he is most famous was an attack on a German machine gun position on October 8, 1918. During the engagement, about half of his squad was killed, and York, a corporal, found himself in command of the remaining seven men. They went on to take the German position as well as 132 German prisoners.

The story of his heroics spread like wildfire, and York was promoted to sergeant. After the war he received an avalanche of decorations and toured the country as a war hero. Later during World War II he served in the signal corps. He died in 1964. After his death his house and several properties he owned were acquired by the state of Tennessee, which incorporated some the sites into a State Park. The Sergeant Alvin C. York State Historic Park was listed on the National Register of Historic Places and named a National Historic Landmark in 1976.

Sergeant Alvin C. York State Historic Park incorporates large open areas, some of which were once part of the York family farm. The main site of interest is the York House, where he lived, which is now home to a museum on York's life. There is also the former York Bible Institute, which York founded with proceeds from a film on his life, and the Wolf River Cemetery, where Alvin York is buried.

128. STONES RIVER NATIONAL BATTLEFIELD

1563 North Thompson Lane, Murfreesboro, Tennessee, 37129

Site Type: Battlefield - Battle of Stones River
Conflict: American Civil War
Dates: Battle fought on December 31, 1862 – January 3, 1863
Designations: National Battlefield
Web: www.nps.gov/stri (official website)

The Battle of Stones River was part of the struggle between the Union and the Confederacy for control of Tennessee in the early years of the war. Following a string of Union victories earlier in 1862, culminating with their victory at Corinth in October, the South sought to retake the initiative in the western theater. A reorganized army under Braxton Bragg, who had won an earlier victory at Murfreesboro, set out to fight the Union Army of the Tennessee.

After months of maneuvering the two armies met at Murfreesboro again. The Union army had a slight numerical advantage, but the Confederacy struck first. On December 31 they launched an attack that inflicted heavy casualties and pushed the Union army back, but failed to break the lines. A series of minor rebel mistakes aided the Union army in remaining intact.

More fighting over the next few days culminated in a major clash on January 3. The Confederacy launched another major assault, this time against a much better prepared Union position. The Confederates were held off with much heavier casualties. While the Confederate army was not defeated, the battle was considered a Union victory. Stones River was designated a National Battlefield in 1927.

Stones River National Battlefield has some of the oldest components of any battlefield in the National Park System. Though it only encompasses a small part of the original battlefield, it includes the remains of one of the largest earthwork forts of the war. Among the numerous monuments and markers commemorating the units that fought in the battle is the Hazen's Brigade Monument, which was erected while the war was still in progress.

129. FRANKLIN BATTLEFIELD

1140 Columbia Avenue, Franklin, Tennessee, 37064

Site Type: Battlefield - Battle of Franklin
Conflict: American Civil War
Dates: Battle fought on November 30, 1864
Designations: National Register of Historic Places, National Historic Landmark District
Web: www.boft.org (official website)

The Battle of Franklin in 1864 was part of a last-ditch effort by the Confederacy to revive the war in Tennessee. A Confederate army under the command of John Hood attempted to take on a vastly superior force by dealing with it in pieces. However the rebels met ferocious opposition from entrenched Union troops and were disastrously defeated. Although the Confederate army remained intact and fought a few more battles, after Franklin they effectively offered no major threat to the North.

Following the fall of Atlanta the Confederate forces in the west were left to their own devices as the Union Army began closing in for the kill in Virginia. Desperate for a victory, or at least a distraction, the last major intact Confederate army in the west moved into Tennessee. Hood was disappointed when he failed to swell his numbers with either reinforcements from the Confederate government or local sympathizers.

Nevertheless he pressed ahead. Seeing an opportunity he attacked the city of Franklin where he was opposed by only a portion of the Union army in Tennessee. However, the well entrenched Northerners held off Hood's army handily, inflicting massive casualties in the process. By the time the Confederates withdrew, they had suffered one of their worst defeats of the war. The Franklin Battlefield was listed on the National Register of Historic Landmarks in 1966 and named a National Historic Landmark District in 1960.

Franklin Battlefield is partially preserved in the Eastern Flank Battlefield Park. Only a small portion survives. This includes several buildings that stand from the time of the war. Some of these buildings still bear the scars of the battle.

---

130. FORT DONELSON NATIONAL BATTLEFIELD

174 National Cemetery Drive, Dover, Tennessee, 37058

Site Type: Battlefield - Battle of Fort Donelson
Conflict: American Civil War
Date: February 11-16, 1862
Designations: National Battlefield
Web: www.nps.gov/fodo (official website)

The Battle of Fort Donelson was arguably the most important victory of the Union Army in the first year of the American Civil War. The battle, really a siege of a Confederate force holed up inside of the fort, resulted in the capitulation of almost an entire rebel army and opened a major gap in the South's strategic defensive position.

In the early months of the war the Union acted quickly to secure the western border states of Missouri and Kentucky. Though successful, their hold was tenuous, so the North followed up almost immediately with operations in Tennessee. In 1862 a Union army under the command of Ulysees S. Grant captured Fort Henry on February 6. Their next target was Fort Donelson.

A large Confederate force was already established at Donelson, but despite reinforcements they were significantly outnumbered. Fighting between the two sides went on for several days and included an unsuccessful breakout attempt by the Confederate army. They were forced to surrender the fort on February 16, and virtually all of the defenders were taken prisoner. Fort Donelson was established as a National Battlefield in 1928.

Fort Donelson National Battlefield encompasses some of the battlefield site as well as a small area of Fort Henry. Portions of the earthworks have been preserved, and there is a battery of cannon overlooking the Cumberland River on display. The battlefield is also home to the Fort Donelson National Cemetery where many of the dead from the engagement are buried.

---

131. SHILOH NATIONAL MILITARY PARK

1055 Pittsburg Landing Road, Shiloh, Tennessee, 38376

Site Type: Battlefield - Battle of Shiloh
Conflict: American Civil War
Dates: Battle fought on April 6-7, 1862
Designations: National Military Park
Web: www.nps.gov/shil (official website)

The Battle of Shiloh was the first truly massive military engagement in the western theater of the American Civil War. It was also one of the most confusing in terms of who won. Tactically it was more or less a draw, but despite Confederate claims to the contrary, Shiloh left the Union army in the western theater in a much more advantageous position than before the battle.

Following the capture of forts Henry and Donelson, Grant led the Army of the Tennessee on a relentless drive southwards intent on capturing the rail hub at Corinth in northern Mississippi. A Confederate army under P.G.T. Beauregard marched north from Corinth to stop them. The two armies met at Pittsburg Landing in southern Tennessee.

The first part of the battle began on April 6 with a Confederate surprise attack that nearly routed the Union army. Only a fierce defense, anchored by the Union soldiers at the Hornet's Nest, kept the Northern army intact. The next day, reinforced by a large army, Grant counterattacked, driving the rebels back from Pittsburg Landing. When the fighting was done both sides had taken heavy casualties, and the Confederacy had fallen back. Shiloh was established as a National Military Park in 1894.

Shiloh National Military Park incorporates a large portion of the

Shiloh Battlefield. Monuments and signage in the park note where certain military units were deployed and where the fighting took place. The park highlight from the perspective of the battle is the Sunken Road where the fiercest fighting at the Hornet's Nest took place.

---

132. NATIONAL WORLD WAR II MUSEUM

945 Magazine Street, New Orleans, Louisiana, 70130

Site Type: Museum
Conflict: World War II
Dates: Opened in 2000
Web: www.nationalww2museum.org (official website)

The National World War II Museum has been officially recognized as such since 2003. Founded in 2000 as the D-Day Museum, it originally focused on the Normandy Invasion of 1944 and the role of New Orleans in that titanic battle. It has since expanded significantly, with exhibits on all of the theaters of the war and America's role in each.

There are eight permanent exhibits at the museum, the oldest and most popular of which is the one concerning the D-Day Invasion of Normandy. In addition to the artifacts from the invasion kept on display here, there is a Higgins boat in the main lobby. The Higgins boats, which carried many of the men and most of the equipment ashore on June 6, were built and tested in the New Orleans area.

Other galleries at the National World War II Museum focus on the histories of the War in Europe and the War in the Pacific, the American home front, the Merchant Marine, and Louisiana's participation in the war. There is also an interactive exhibit on the USS Tang, the most successful American submarine of the war. Other artifacts on display include an Enigma coding machine, a Sherman Tank, and an assortment of aircraft and artillery pieces.

133. JEAN LAFITTE NATIONAL HISTORICAL PARK

8606 West St. Bernard Highway, Chalmette, Louisiana, 70043

Site Type: Battlefield – Battle of New Orleans
Conflict: War of 1812
Dates: Battle fought on January 8, 1815
Designations: National Register of Historic Places, National Historic Park and Preserve
Web: www.nps.gov/jela/chalmette-battlefield.htm (official website)

The Battle of New Orleans was the final military engagement of the War of 1812, famously taking place after the war had technically ended as the belligerents had not yet been informed of the cessation of hostilities. One of the larger battles that had yet taken place on American soil, it involved a ragtag force of soldiers, sailors, militia, armed civilians, slaves and even pirates preventing a superior British army from capturing the city of New Orleans. It was America's largest, if superfluous, victory of the war.

By late 1814 the war had been dragging on for over two years. After stalemates in the north, American victory in the west and the burning of Washington by the British, both sides were ready for an end to hostilities. However, an effort to capture New Orleans was already underway and being planned in the Caribbean. Even as the Treaty of Ghent was being signed in Europe in December of 1814, a huge British invasion force was being dispatched towards Louisiana.

Future president Andrew Jackson was put in charge of the defense, and he hastily scraped together every able body he could muster. By early January the American force was entrenched before the city. A British amphibious assault attacked the position on January 8 and in a short time was repulsed with massive casualties. Over two thousand British soldiers were killed, wounded or captured, compared to only a few hundred Americans. By the end of the month the British had abandoned the campaign, just before the Treaty of Ghent was ratified by the U.S. Congress. The battlefield site was listed on the National

Register of Historic Places in 1966 and incorporated into the Jean Lafitte National Historic Park and Preserve in 1978.

The Jean Lafitte National Historic Park is named after the famous French privateer who came to the aid of the American defense in exchange for a pardon for his crew. The park incorporates the old Chalmette Battlefield and National Cemetery where the battle actually took place, as well as an extensive protected wetland area and preserve. Some of the earthworks from the battle have been preserved, while a massive obelisk commemorates the victory.

---

134. FORT SMITH NATIONAL HISTORIC SITE

301 Parker Avenue, Fort Smith, Arkansas, 72901

Site Type: Military Base
Conflicts: Early American Indian Wars, American Civil War
Dates: Originally completed in 1838
Designations: National Register of Historic Places, National Historic Site
Web: www.nps.gov/fosm (official website)

Fort Smith is one of the oldest and best preserved forts in the southern part of the early western frontier. Originally established in 1817, it was abandoned, reoccupied and rebuilt several times, including by the Confederacy during the Civil War. Although not used since the 19th century, much of the fort is intact, a rarity for the time period and region.

The area where Fort Smith now stands has been militarily important since the time of the Louisiana Purchase. An original wooden stockade fort stood here from 1817 to 1824, but was abandoned in favor of another site before being reconstructed in 1838. Fort Smith became an important way station along the infamous Trail of Tears when the Cherokee tribes were forcibly relocated westward. It was also used as an operating base during the Mexican-American War.

Fort Smith was occupied by the Confederacy at the outset of

hostilities during the Civil War, but recaptured by the Union in 1863. Fort Smith's occupation by the Union helped to permanently secure Missouri for the North. The fort was decommissioned in 1781. The Fort Smith National Historic Site was established in 1960 and listed on the National Register of Historic Places in 1966.

A number of important buildings are preserved here, including the barracks, which later housed the courthouse presided over by famous hanging judge Isaac Parker and which now houses the visitor's center. Some of the foundations of the original fort buildings are here, as well as historic markers noting places associated with the Trail of Tears.

---

135. PEA RIDGE NATIONAL MILITARY PARK

15930 East Highway 62, Garfield, Arkansas, 72732

Site Type: Battlefield – Battle of Pea Ridge
Conflict: American Civil War
Dates: Battle fought on March 7-8, 1862
Designations: National Register of Historic Places, National Military Park
Web: www.nps.gov/peri (official website)

The Battle of Pea Ridge was one of the most ambitious, if unsuccessful, Confederate counteroffensives in the western theater of the American Civil War. Launched early in 1862, it represented the South's last major effort to wrest Missouri from Northern control. The failure of this campaign helped to secure the northwestern frontier for the Union for the rest of the war.

Throughout the latter months of 1861 the Union army succeeded in driving the Missouri State Guard, composed of pro-slavery sympathizers, southward into Arkansas. In early 1862 the Union invaded Arkansas where a Confederate force was being assembled to stop them. The North, recognizing that it was in danger of becoming overextended, fell back and fortified a position near Leetown, Arkansas.

Warned of the approaching Confederate counterattack the Federal soldiers were well prepared. In two days of fighting the rebels assaulted the Union positions, and casualties were heavy on both sides. However, the Northern army could not be dislodged, and the Confederates were forced to fall back. Pea Ridge was named a National Military Park in 1856 and listed on the National Register of Historic Places in 1966.

Pea Ridge National Military Park incorporates the entire site of the battle, including the Elkhorn Tavern, where some of the fiercest fighting raged while the proprietor and his family waited out the battle in the building's cellar. All of the other original buildings in the area have long since been demolished. Also in the park is a section of the infamous Trail of Tears, which is commemorated by markers and a designated walking path.

GREAT PLAINS

136. FORT SNELLING

200 Tower Avenue, St. Paul, Minnesota, 55111

Site Type: Masonry Fort
Dates: Originally completed in 1825
Designations: National Register of Historic Places, National Historic Landmark
Web: www.historicfortsnelling.org (official website)

Fort Snelling is one of the oldest fortifications standing in the northern Great Plains. Begun in the 1810s, Fort Snelling, originally named Fort Anthony, is a large stone-built fortress rather than a typical frontier-style wooden stockade fort. Because of this the fort is well preserved and remained in use all the way through World War II.

Construction on Fort Anthony began in 1819, primarily as a defense against a possible invasion from Canada following the War of 1812. It was completed in 1825 at which time it was renamed Fort Snelling. It became an early trading post for settlers migrating into Minnesota and beyond in the years leading up to the American Civil War.

Troops from Snelling fought in most American conflicts in the second half of the 19th century. The fort was used as a language training center during World War II and remained an active military base until 1946. Fort Snelling was named a National Historic Landmark in 1960 and listed on the National Register of Historic Places in 1966.

Fort Snelling is fairly intact, with most of the parade ground and surrounding buildings still standing. The fort is anchored by a massive round tower which, while constructed to serve as a platform for artillery, looks more like a medieval European castle. A memorial at the fort commemorates those Native Americans who were once imprisoned here.

137. FORT ATKINSON STATE PRESERVE

303 2nd Street NW, Fort Atkinson, Iowa, 52144

Site Type: Masonry and Wood Fort
Dates: Originally completed in 1842
Designations: National Register of Historic Places, National Historic District, Iowa State Preserve
Web: www.iowadnr.gov/places-to-go/state-preserves (official website)

Fort Atkinson was one of the earlier major army forts established on what was then America's northwestern frontier. Built to help establish a buffer between mutually hostile tribes in the area as well as to prevent the trespassing of white settlers, the fort remained in active service for less than a decade. For most of that time it helped to keep control over the tribes that had been forcibly relocated westward from Wisconsin.

In the 1830s the United States government facilitated a peace treaty with a number of Native American tribes on the Great Plains frontier. The treaty established a large neutral zone separating the local Fox and Sioux. Not too long afterwards the relatively small neutral zone was redesignated as a resettlement area for Ho-Chunk Indians who were being forced to leave their homes in Wisconsin.

In 1840 the army sent a detachment of troops to the area, and two years later they had constructed Fort Atkinson. Most of the troops left the fort when hostilities broke out with Mexico in 1846. After that it was only periodically occupied and ultimately abandoned in 1849. Fort Atkinson was made a State Reserve in 1968 and listed on the National Register of Historic Place in 2013.

Fort Atkinson State Preserve is a partially restored, partially recreated complex giving a fairly good idea of what the fort originally looked like. Surviving buildings include several barracks, the guard house and the powder magazine surrounding a large, open parade ground. The fort hosts an annual recreation of life on the frontier in the mid-19th century.

138. STARS AND STRIPES MUSEUM AND LIBRARY

17377 Stars and Stripes Way, Bloomfield, Missouri, 63825

Site Type: Museum
Dates: Opened in 1991
Web: www.starsandstripesmuseumlibrary.org (official website)

The Stars and Stripes Museum and Library is the archive and gallery of *The Stars and Stripes*, the official newspaper of the United States military. Opened in 1991, it houses exhibits on this iconic news source from the Civil War to the present day. The museum is part of the Stars and Stripes Historical and Cultural Byway.

The Stars and Stripes newspaper was first published by Union troops during the early months of the American Civil War. It is believed that the first edition was printed in an abandoned newspaper office on or around November 9, 1861. It was not maintained for a lengthy period, but it apparently sparked interest in the idea for a military newspaper.

The Stars and Stripes was published during both World Wars, and by 1945 had become a cultural icon. After World War II, the newspaper was maintained for the troops that remained stationed overseas. It has continued to be regularly published ever since. It remains a widely read news source, with a reported seven million editions distributed in 2019.

The Stars and Stripes Museum and Library was created in Bloomfield, the town where the first papers were printed over a century and a half ago. The majority of the collection consists of a chronological series of exhibits covering the time from the 1860s until the present. The library and archive houses the largest known collection of *The Stars and Stripes* editions, as well as many books and artifacts related to the history of the military newspaper.

139. WILSON'S CREEK NATIONAL BATTLEFIELD

6424 West Farm Road 182, Republic, Missouri, 65738

Site Type: Battlefield - Battle of Wilson's Creek
Conflict: American Civil War
Dates: Battle fought on August 10, 1861
Designations: National Register of Historic Places, National Battlefield
Web: www.nps.gov/wicr (official website)

The Battle of Wilson's Creek was the first large battle of the American Civil War west of the Mississippi. Fought between federal occupying forces and state militia sympathetic to the Confederacy, the battle was a victory of sorts for the South. However due to their heavy casualties the Confederates were unable to follow-up their victory, and Missouri was annexed by the Union shortly thereafter.

In the early months of the Civil War, as states across the South seceded one after another, Missouri initially attempted to maintain a state of neutrality in the conflict. However many elements, including much of the government, were clearly sympathetic to the Confederate cause, and by the late summer the North sent in an army in order to secure Missouri's place in the Union.

In August a Federal force was sent to crush the Missouri State Guard. The two armies met at Wilson's Creek on August 10th, with the Union gaining an early battlefield advantage. However, problems with communications and several devastating counterattacks led to a Union retreat. Despite their victory the Confederates could not sustain their losses and were forced to pull back. Wilson's Creek was declared a National Battlefield Park in 1960 and listed on the National Register of Historic Places in 1966.

Wilson's Creek National Battlefield Park encompasses most of the original battle site, including Bloody Hill, where some of the most intense fighting occurred. The visitor's center is home to the Wilson's

Creek Civil War Museum, which has exhibits and artifacts from campaigning that took place all along throughout the western theater. Also on site is the Hulston Library which has a collection of books devoted to the Civil War west of the Mississippi.

---

140. NATIONAL WORLD WAR I MUSEUM AND MEMORIAL

2 Memorial Drive, Kansas City, Missouri, 64108

Site Type: Monument – World War I
Conflict: World War I
Dates: Dedicated in 1926
Web: www.theworldwar.org (official website)

The National World War I Museum and Memorial has been officially recognized as such by the United States government since 2004. Founded in 1926 as the Liberty Memorial, it was originally a monument to commemorate those who served during the Great War. It has since been significantly expanded to include a large museum with exhibits on the war both prior to and after the American entry into the conflict.

The First World War was one of the largest foreign conflicts that the United States has ever been involved in, and the third deadliest in terms of casualties. Millions served in uniform, along with forces from virtually every major world power at the time. The modern American military as we know it largely grew out of this massive conflict.

The National World War I Museum and Memorial in Kansas City has an Egyptian theme. The centerpiece and main monument is the Liberty Memorial, an immense obelisk-like tower flanked by a pair of enormous sphinxes. The site is home to a sizeable museum that houses a large collection of military artifacts, representing both the Allied and enemy nations. Many items of interest can be found here, from one of the war's last surviving tanks to personal items owned by John J. Pershing and even Paul von Hindenberg.

141. FORT ABERCROMBIE STATE HISTORIC SITE

County Highway 22, Abercrombie, North Dakota, 58001

Site Type: Wooden Stockade Fort
Conflict: Later American Indian Wars
Dates: Originally completed in 1860
Designations: National Register of Historic Places, North Dakota State Historic Site
Web: www.history.nd.gov/historicsites/abercrombie (official website)

Fort Abercrombie was one of the first military outposts to be established in the Dakota Territory. It was also one of only a handful of forts to see action against Native American tribes in the region during the Civil War. Largely abandoned in the 1870s, most of the fort as it stands today is a recreation. The site is also home to a museum.

In the years leading up to the American Civil War, the Dakota Territory was starting to see increasing activity from settlers and traders. Fort Abercrombie was founded in 1858 to protect the area, with the current site of the fort settled in 1860. In 1862 the area was attacked by the Sioux, and the fort besieged unsuccessfully for over a month.

After the Civil War Fort Abercrombie became an important transportation hub, protecting local military traffic as well as the newly arrived railroad. It was abandoned in 1877 and largely dismantled. Much of the site was restored in the late 1930s by the Works Progress Administration. Fort Abercrombie was listed on the National Register of Historic Places in 2009.

Fort Abercrombie is mostly a recreation, but there are some original elements including the guardhouse. The blockhouses and palisade are very good reconstructions. A small museum on the site has a collection of artifacts and exhibits related to the history of the fort.

142. MINUTEMAN MISSILE NATIONAL HISTORIC SITE

24545 CottonWood Road, Philip, South Dakota, 57567

Site Type: Missile Silo
Conflict: Cold War
Dates: Completed in 1963
Designations: National Register of Historic Places, National Historic Site
Web: www.nps.gov/mimi (official website)

Minuteman Missile National Historic Site recalls a time in American history when nuclear war between super powers was a frighteningly real possibility. Located inside an actual nuclear missile launch facility, and featuring an actual though non-functional Minuteman Missile, this fearsome reminder of the Cold War is one of the most unique and sobering places of military interest in the United States.

In 1945 America tested its first atomic bomb. Four years later, the Soviet Union tested its first atomic bomb. For the next four decades, these two super powers and several other nations created enormous nuclear arsenals with the power to make the planet Earth uninhabitable. At its peak in the 1960s and 1970s the United States had more than thirty thousand nuclear weapons.

Fortunately, despite several close calls, no other nuclear weapons have been used against an enemy target since the end of World War II. Following the Cold War the American arsenal was greatly reduced, and the Minuteman II ICBM program completely eliminated. Today only one remains, preserved at the Minuteman Missile National Historic Site. This last surviving Minuteman Missile silo was named a National Historic Site and listed on the National Register of Historic Places in 1999.

Minuteman Missile National Historic Site is one of the most haunting destinations in the entire National Park system. On the surface only a small visitor facility marks its location, as it was largely disguised in the 1960s. Visitors can peer down into the actual silo where the last surviving Minuteman II missile maintains its impotent vigilance. The site also has exhibits on the Cold War and the Arms Race, both of which defined several generations of Americans.

---

143. CRAZY HORSE MEMORIAL

12151 Avenue of the Chiefs, Crazy Horse, South Dakota, 57730

Site Type: Monument
Conflict: Later American Indian Wars
Dates: Dedicated in 1998 (construction currently in progress)
Web: www.crazyhorsememorial.org (official website)

The Crazy Horse Memorial is one of the largest monuments ever conceived. Designed to honor one of the greatest Native American war chiefs to have ever lived, this enormous mountain carving was begun in 1948 and is still a work in progress more than seven decades later.

Crazy Horse was one of the last and greatest leaders of the Lakota tribes in the latter half of the 19th century. In the wake of the American Civil War, thousands of settlers began moving west into the Great Plains, the last real stronghold of Native American territory in the continental United States. In desperation, leaders such as Crazy Horse rallied the People to a valiant albeit futile defense against settler encroachments.

From 1866 until 1877 Crazy Horse masterminded a brilliant guerilla war against the American Army. His greatest victory came at the Battle of Little Bighorn in 1876, where his forces destroyed a large detachment of American cavalry under the command of George Custer. Once it became clear that his war was unwinnable, Crazy Horse surrendered in 1877. He was murdered in prison a few months later.

The Crazy Horse Memorial was conceived of by Henry Standing Bear, a Lakota chief, in the mid-20th century. The monument was meant to overshadow the nearby presidential carving at Mount Rushmore, which was seen by many Native American groups across the

United States as a great insult due to the sacred nature of the Black Hills. Although it has been under construction for more than seventy years, to date only the face and part of an outstretched arm of the monument have been completed.

---

144. FORT ROBINSON STATE PARK

3200 Highway 20, Crawford, Nebraska, 69339

Site Type: Military Base
Conflict: Later American Indian Wars
Dates: Established in 1876
Designations: National Register of Historic Places, National Historic District
Web: www.outdoornebraska.gov/fortrobinson (official website)

Fort Robinson is one of the more storied American frontier forts in the Great Plains. It was known for its horse and cavalry training, as well as being the site of the death of great Native American chief Crazy Horse and the Fort Robinson Massacre. It was later active as a base for the Buffalo Soldier regiments, a K-9 training center and as a POW camp during World War II.

During the 1870s large numbers of settlers began heading west into the Great Plains, often with support from the United States Army. In 1873 thousands of Lakota tribes-people were resettled in Nebraska, and Fort Robinson was constructed to ensure there were no problems. The garrison of the fort was very active during the Sioux Wars, and after Crazy Horse was captured he was brought to Fort Robinson where he was later executed. In 1879 nearly a hundred Cheyenne were killed or wounded here during the Fort Robinson Massacre.

By the 20th century Robinson was being used as a training center. The fort was greatly expanded in the early 20th century and used for the training of horses and dogs for the army. German POW's were interred here during the Second World War. The fort was decommissioned in 1947, and a museum opened on the site in 1956. Fort

Robinson was established as a National Historic Landmark District in 1960 listed on the National Register of Historic Places in 1966.

Fort Robinson is a largely intact military site thanks in large part to its continued use through the middle of the 20th century. The main point of interest is the Post Headquarters which now houses the Fort Robinson Museum. The museum has exhibits on the fort's extensive history, notably its work with the Lakota Indians in the 1870s. A plaque marks the location where Crazy Horse died.

---

145. FORT LEAVENWORTH

1 Sherman Avenue, Fort Leavenworth, Kansas, 66027

Site Type: Military Base
Dates: Established in 1827
Designations: National Historic Landmark District
Web: https://home.army.mil/leavenworth (official website)

Fort Leavenworth is one of the oldest active military posts in the United States Army. Founded in 1827, it became one of the major transit points for travelers working their way into and around the west. In addition to being home to the United States Army Combined Arms Center, Fort Leavenworth is famous as the location of the Disciplinary Barracks, America's maximum security military prison.

The site where Leavenworth now stands has been an important way station since the French arrived in the area in the 18th century. Lewis and Clark came through here in 1804. A military post was established in 1827 to protect travelers on the Santa Fe Trail. Leavenworth was a critical operating base in the Mexican-American War, the Civil War and various American Indian wars throughout the 19th century.

In 1881 William T. Sherman established a training center at Leavenworth that went on to become the Army Command and General Staff College. In 1874 Leavenworth's most famous institution, the United States Disciplinary Barracks, was opened. This facility is still used for the incarceration and occasional execution of military prisoners. Parts of Fort Leavenworth were included in the Freedom's Frontier Heritage National Historic Landmark District in 1974.

Fort Leavenworth is a fully functional and very large military base. Sites of interest include the Frontier Army Museum, with exhibits on the fort's history, the United States Disciplinary Barracks and the Fort Leavenworth National Cemetery. A huge bronze statue of a mounted Buffalo Soldier commemorates their service at Leavenworth. While the installation can be toured, Fort Leavenworth is an active military base, and visits are likely to be subject to restrictions.

---

146. BLACK JACK BATTLEFIELD AND NATURE PARK

163 East 2000 Road, Wellsville, Kansas, 66092

Site Type: Battlefield – Battle of Black Jack
Conflict: Bleeding Kansas
Dates: Battle fought on June 2, 1856
Designations: National Register of Historic Places, National Historic Landmark
Web: www.blackjackbattlefield.org (official website)

The Battle of Black Jack was a minor but famous paramilitary engagement that took place during the turbulent conflict known as Bleeding Kansas. Fought between two relatively small groups of militia, in which there were few casualties, Black Jack was at best a skirmish, but one with important consequences as it elevated the Kansas conflict and brought America one step closer to civil war.

From 1855 through 1858 Kansas experienced a turbulent period where pro- and anti-slavery militias vied for control of the territory. Not really a war as much as a blood feud, sporadic fighting over the course of three years left around sixty killed and another hundred hurt. The most famous figure of the conflict was undoubtedly New England preacher and rabid abolitionist, John Brown.

Brown fought at several battles, but his most successful was Black

Jack. In the wake of two of his sons being captured by pro-slavery militia, Brown led a force of about thirty anti-slavery militiamen against a roughly equal number of the enemy. After several hours of fighting the abolitionists compelled the pro-slavery faction to surrender. Brown agreed to let his enemies go in exchange for his sons. News of the engagement rallied both sides to widen the conflict. The Black Jack Battlefield was listed on the National Register of Historic Places and named a National Historic Landmark in 2012.

The Black Jack Battlefield preserves part of the area where the skirmishing took place. The centerpiece of the historic site is the Robert Hall Pearson House, which still stands from the time of the battle. As of the time of this writing the park was in the process of expanding its facilities and signage.

---

147. FORT SCOTT NATIONAL HISTORIC SITE

Old Fort Boulevard, Fort Scott, Kansas, 66701

Site Type: Military Base
Conflict: Bleeding Kansas
Dates: Mostly completed by 1850 (never finished)
Designations: National Register of Historic Places,
National Historic Site, National Historic Landmark District
Web: www.nps.gov/fosc (official website)

Fort Scott was one of the shorter-lived of the major frontier army bases of the 19th century, though it did play a notable role in the events of the Bleeding Kansas conflict. Only partially completed by 1850, the fort was briefly occupied, then abandoned, then re-occupied, then abandoned a final time in 1873. The historic site now includes both the surviving buildings and a small protected wilderness area.

Fort Scott was one of several bases primarily used by the cavalry that protected the frontier in the years leading up to the Civil War. Considered obsolete even before its completion, work on the fort was stopped in 1850. A few years later its buildings were taken over by pro- and anti-slavery factions vying for political control of the territory.

As a result several violent incidents took place here. In 1861 the army re-occupied the fort to prevent the area from coming under the influence of the Confederacy. After the Civil War the fort was abandoned. It was reactivated for a brief period in the early 1870s as a base to protect the railroads. The army moved out in 1873. Fort Scott was listed on the National Register of Historic Places in 1966 and designated a National Historic Site in 1978.

Most of the original structures of Fort Scott are gone. However a few buildings survive, including several barracks, stables, powder magazine and the hospital. About a third of the park is given over to a small field of wild prairie grass.

---

148. DWIGHT D. EISENHOWER PRESIDENTIAL LIBRARY AND MUSEUM

200 South East 4th Street, Abilene, Kansas, 67410

Site Type: Museum
Dates: Opened in 1962
Web: www.archives.gov/presidential-libraries/visit/eisenhower (official website)

The Dwight D. Eisenhower Presidential Library is located in the general's hometown of Abilene. Unlike most other presidential libraries, this one began to take shape before Eisenhower even became president. In honor of his long and distinguished service in the United States Army, work on a museum honoring his military career began in the early 1950s. It opened in 1954, a little over a year into his presidency.

By 1959 it was decided to expand the facility to include Eisenhower's presidential library. It officially opened in 1962, a year after the end of his second term in office. Eisenhower was buried at the site after his death in 1969.

The Dwight D. Eisenhower Library and Museum now includes the presidential library, the museum on his life, his boyhood home and a meditation chapel where he and his wife are buried. Most of the artifacts and papers related to his military career can be found in the museum. At the time of this writing, plans to renovate and expand the museum were in progress.

---

149. FORT RILEY MUSEUMS

885 Henry Avenue, Fort Riley, Kansas, 66442

Site Type: Military Base, Museums
Dates: Base established in 1853
Web: https://home.army.mil/riley/index.php/about/museums (official website)

The Fort Riley museums are a collection of museums and historic sites located on the grounds of Fort Riley, one of the largest American military bases in the Great Plains. Founded in 1853, it was an important center of cavalry activity and training in the second half of the 19th century. In later years it served as a base for the Buffalo Soldiers and the 1st Infantry Division among other units.

The most famous person associated with Fort Riley is probably George Armstrong Custer. Custer was in command of the fort in the years following the Civil War, and it was here that the 7th cavalry was stationed. His home is still preserved on the base. The fort is now home to the United States Cavalry Museum, which houses exhibits from the 18th through the 20th centuries, as well as the Custer House. Both of these sites are popular with cavalry buffs.

Fort Riley is also home to several other museums, including the First Infantry Division Museum, which chronicles their one hundred year history; and the First Territorial Capitol of Kansas Museum, which chronicles the history of the Kansas Territory. Fort Riley remains an active military base and visitors may be subject to restricted access.

150. FORT LARNED NATIONAL HISTORIC SITE

1767 KS Highway 156, Larned, Kansas, 67550

Site Type: Military Base
Conflict: Later American Indian Wars
Dates: Originally completed in 1860
Designations: National Register of Historic Places, National Historic District, National Historic Site
Web: www.nps.gov/fols (official website)

Fort Larned was one of the key military installations set up by the army to protect the Santa Fe Trail in Kansas in the mid-19th century. The fort saw a lot of activity during the American Civil War, and afterwards was used by George Custer in his campaigns against the Plains tribes. The site is fairly intact thanks to the stone construction used by many of its buildings.

Throughout the 1850s commerce along the Santa Fe Trail had grown significantly, and it was decided a new military base was required to support and protect travelers on the road. An initial camp was established in 1859, followed by an adobe fort a few years later. During the Civil War, when the professional garrison was replaced by volunteer militia, Native American raids on the fort and the surrounding area increased. In 1862 the Confederacy planned to take Fort Larned with the help of local tribes, but the plan was abandoned.

After the Civil War Fort Larned became a major base of operations for the United States Army. The fort was re-garrisoned and rebuilt. In 1867 George Custer led a force from here to burn down a Cheyenne village. A year later Philip Sheridan used Fort Larned as a base in his efforts to drive the local tribes onto reservations. The fort remained active until 1878, after which it was used for ranching. Fort Larned became a National Historic Site in 1964 and was listed on the National Register of Historic Places in 1966.

Fort Larned National Historic Site is a large, well preserved installation thanks both to its superior construction and its ongoing use. Many of the original buildings are intact and have been restored, including both enlisted and officer barracks, the blockhouse, the commissary and various commercial buildings and stores. Most of these surround the sprawling parade grounds. There is also a cemetery on site.

---

151. FORT SILL & UNITED STATES ARMY ARTILLERY MUSEUM

Corral Road, Fort Sill, Oklahoma, 73503

Site Type: Military Base, Museum
Dates: Base established in 1869
Designations: National Register of Historic Places, National Historic Landmark
Web: https://sill-www.army.mil/famuseum (official website)

Fort Sill is one of the most enduringly historic forts of the American frontier. Still an active military base, it is home to the United States Army Field Artillery School, Air Defense Artillery School and several active artillery Brigades. Because of its close association with the use of artillery, the Field Artillery Museum was established here in 2009.

Fort Sill was established as a base to help secure the Oklahoma Territory during the Later American Indian Wars. Its commanders during this period included Philip Sheridan and William Sherman of Civil War fame. Scouts based here included both Wild Bill Hickock and Buffalo Bill Cody. Famous Native American war leader Geronimo was incarcerated here in the 1890s.

In the 20th century Fort Sill became a primary training center for field artillery units. It was also the site of one of the earliest military airfields, with pilots trained here for duty in World War I. During World War II both suspected Japanese spies and German POW's were incarcerated here. Fort Sill was named a National Historic Landmark in 1960 and listed on the National Register of Historic Places in 1966.

Fort Sill is home to several museums, including one encompassing most of the original fort's historic buildings. The U.S. Army Artillery

Museum has one of the most comprehensive collections of artillery, including over seventy large guns. The fort's cemetery has the remains of not only soldiers and their families, but important Native American leaders as well, including the renowned chief Geronimo.

---

152. WASHITA BATTLEFIELD NATIONAL HISTORIC SITE

18555 Highway 47A, Cheyenne, Oklahoma, 73628

Site Type: Battlefield – Battle of Washita
Conflict: Later American Indian Wars
Dates: Battle fought on November 27, 1868
Designations: National Register of Historic Places, National Historic Landmark, National Historic Site
Web: www.nps.gov/waba (official website)

The Battle of Washita River, also known as the Washita Massacre, was an attack by federal troops led by George Custer on a Cheyenne encampment in the Oklahoma Territory. Hailed at the time as an important American victory against aggressive native tribes along the frontier, it is recognized in hindsight as being better described as a massacre that involved the killing of many innocent non-combatants. The battlefield is now commemorated as such.

In 1868, after having received an order requiring Native Americans in Kansas and Colorado to relocate to Oklahoma, various tribes, including the Arapahoe, Comanche, Pawnee and Cheyenne began to organize and push back. Throughout the summer raids were carried out against white settlements, raising the ire of the federal government. It was determined that some of the raiders came from a camp of Southern Cheyenne near the Washita River.

A large cavalry force was sent out to deal with the Cheyenne. Despite warnings from scouts the tribal leaders discounted the threat and were caught unprepared when the United States army showed up on November 27. The result was a horrific defeat for the Cheyenne.

Though estimates differ, most of the warriors were killed, as well as an unknown number of women and children. Washita Battlefield was named a National Historic Landmark in 1965 and listed on the National Register of Historic Places in 1966.

Washita Battlefield National Historic Site, like many of the other battle sites of the western frontier, is now more about educating visitors about the mistreatment of Native Americans. The site incorporates part of the encampment areas along the Washita River. A lone teepee rather than a stone monument stands in front of the visitor center.

SOUTHWEST

153. BUFFALO SOLDIERS NATIONAL MUSEUM

3816 Caroline Street, Houston, Texas, 77004

Site Type: Museum
Dates: Opened in 2001
Web: www.buffalosoldiermuseum.com (official website)

The regiments that comprised the force known as the Buffalo Soldiers are among the most storied fighting units of the United States Army. Formed in the years after the Civil War, these regiments consisted entirely of African American infantry and cavalrymen who distinguished themselves in almost all of America's major foreign conflicts from the Spanish-American War through World War II. They were also very active on the frontier in the late 19th and early 20th centuries.

The first Buffalo Soldier regiment, the 10th Cavalry, was formed at Fort Leavenworth in 1866. They were later followed by four infantry regiments and an additional cavalry regiment. For nearly a century these soldiers served with bravery and distinction, all while enduring segregation, racism and discrimination. Over two dozen Buffalo Soldiers went on to receive Congressional Medals of Honor during their active years, which ended in 1951. The Buffalo Soldiers National Museum was established in Houston, Texas in 2001.

The Buffalo Soldiers National Museum is housed in the former Houston Light Guard Armory. It is home to a diverse collection of artifacts and gear from virtually every period that the Buffalo Soldier regiments were active, notably the two World Wars. There are also special exhibits on African American soldiers who served in the Civil War prior to the formation of the Buffalo Soldier units, and the Vietnam War after they were disbanded.

154. SAN JACINTO BATTLEGROUND STATE HISTORIC SITE

3523 Independence Parkway South, La Porte, Texas, 77571

Site Type: Battlefield – Battle of San Jacinto
Conflict: Texas Revolution
Dates: Battle fought on April 21, 1836;
Monument dedicated in 1939
Designations: National Register of Historic Places,
National Historic Landmark
Web: https://thc.texas.gov/historic-sites/san-jacinto-battleground-state-historic-site (official website)

The Battle of San Jacinto was the decisive engagement of the Texas revolt against Mexico. Fought in the wake of the defeat at the Alamo, San Jacinto was the turning point of the war. Fought between a large, well-equipped Mexican army and a much smaller force of hastily gathered volunteer militia, the unexpected victory led directly to the independence of Texas.

By April 1836 the Texas Revolution had already been raging for six months, with bloody engagements fought at Goliad and the Alamo. The latter was a pyrrhic victory for the Mexicans in which they were stalled for nearly two weeks outside of San Antonio. This delaying action gave the Texas general Sam Houston time to recruit and prepare fresh forces to stop the Mexican invasion. After the Alamo, Santa Anna attempted to capture the Texas government at Lynch's Ferry, buying Houston even more time. It wasn't until late in the month that the Mexican army turned its full attention to crushing the Texas rebels.

By that time Houston's forces had established a strong position near the San Jacinto River. The Mexicans prepared to attack on the 22nd, but the Texans pre-empted this by launching their own attack the day before, taking the enemy completely by surprise. Santa Anna's army suffered massive casualties, while Texas lost only nine killed and thirty wounded. The rest of the Mexican army was scattered and forced to withdraw to the Rio Grande River, resulting in a victory for the rebels. The San Jacinto Battlefield State Historic Site was named a National Historic Landmark in 1960 and added the National Register of Historic Places in 1966.

San Jacinto Battlefield State Historic Site encompasses a large area in La Porte, including most of the area where the battle took place. It is dominated by the San Jacinto Monument, a 567-foot tall obelisk-like tower crowned by a massive star. The base of the monument houses the San Jacinto Museum of History, which features exhibits on the battle and on early Texas history.

---

155. BATTLESHIP TEXAS STATE HISTORIC SITE

3523 Independence Parkway South, La Porte, Texas, 77571

Site Type: Battleship
Conflicts: World War I, World War II
Dates: Commissioned in 1914
Designations: National Register of Historic Places, National Historic Landmark
Web: https://tpwd.texas.gov/state-parks/battleship-texas (official website)

The USS Texas (BB-35) is the oldest surviving battleship of the dreadnaught era. Commissioned in 1914, it is also the only remaining American battleship that saw action in World War I. Between the wars it served as the American flagship and was used for presidential escort duties in the Atlantic. One of the few American battleships that was not at Pearl Harbor when the Japanese attacked in 1941, the Texas was one of the first in action, and was present at several of the largest naval engagements of World War II.

The Texas was commissioned on the eve of the First World War. Her first duty was patrolling in the Gulf of Mexico during a tense period between Mexico and the United States. When America entered the war, the Texas fired the first American shots, specifically at a German U-boat. The Texas was largely relegated to convoy duty for most of the war and spent the interwar years running routine assignments in the Atlantic.

World War II saw the USS Texas as one of the most important American capital ships active in the European Theater. The Texas participated in the American invasion of North Africa in 1942, with Walter Cronkite reporting from the ship during the engagement. On D-Day the Texas supported American landing forces at Omaha Beach and Pointe du Hoc. She then participated in the attack on Cherbourg and later in the landings in Southern France. The Texas then made it to the Pacific in time to participate in the attacks on Iwo Jima and Okinawa. After the war America's oldest battleship was decommissioned in 1948. The USS Texas was added to the National Register of Historic Places and designated a National Historic Landmark in 1976.

The USS Texas was acquired by the state of Texas after the war and refurbished as America's first battleship museum. It was also the first battleship to be declared a National Monument. In addition to being one of the largest battleships constructed prior to World War II, the Texas was also one of the first ships to boast most of the technological naval advances of the mid-20th century. It is currently moored in the Houston Ship Channel where it is part of the San Jacinto Battlefield State Historic Site.

---

156. USS LEXINGTON

2914 North Shoreline Boulevard, Corpus Christi, Texas, 78402

Site Type: Aircraft Carrier
Conflicts: World War II, Cold War
Dates: Commissioned in 1943
Designations: National Register of Historic Places, National Historic Landmark
Web: www.usslexington.com (official website)

The USS Lexington (CV-16), the *Blue Ghost*, was the longest-serving aircraft carrier in American history. It was named in honor of its predecessor which was sunk at the Battle of the Coral Sea. The Lexington fought in several major engagements of the Pacific war, including the Battle of Leyte Gulf and the Marianas Turkey Shoot. She earned the nickname Blue Ghost because of its color and its apparent invincibility as it was thought by the Japanese to have been destroyed on several occasions.

In May 1942 the first USS Lexington aircraft carrier (CV-2) was sunk following the Battle of the Coral Sea. A new carrier then under construction, the Cabot, was renamed and commissioned as the Lexington in February 1943. The new Lexington was rushed into service in the Pacific Theater. Aircraft from the Lexington participated in raids against the Japanese home islands prior to Japan's surrender in 1945.

The Lexington remained in service long after World War II and was only finally decommissioned in 1991, nearly fifty years after its completion. For most of its history it was used as a training carrier in the Gulf of Mexico. A few years later the Lexington was permanently moored at Corpus Christi as a museum ship. The USS Lexington was added to the National Register of Historic Places and named a National Historic Landmark in 2003.

The USS Lexington Museum on the Bay is now home to the ship, which at the time of this writing was undergoing restorations. Every so often a new section is opened to the public. Among the things to see here is a permanent exhibit on the attack on Pearl Harbor, as well as an enormous collection of scale models, with nearly four hundred model ships and planes on display.

---

157. COMMEMORATIVE AIR FORCE AIRPOWER MUSEUM

9600 Wright Drive, Midland, Texas, 79706

Site Type: Museum
Conflict: World War II
Dates: Opened in 1965
Web: https://commemorativeairforce.org (official website)

The Commemorative Air Force Airpower Museum is dedicated to American military aviation during World War II. Most of the aircraft and artifacts on display here are from that conflict. The CAF traces its roots back to 1957 when its founders acquired an old P-51 fighter. This turned out to be the first of tens of thousands of pieces in the collection, including over a hundred aircraft, many of which are still serviceable.

The main items of interest here are the aircraft, which include both American and foreign planes, all dating from World War II. The diverse collection runs from small aircraft to enormous Super Fortress and Flying Fortress bombers. The fleet is collectively known as the Ghost Squadron, and many of these fly at Commemorative Air Force air shows.

In addition to the aircraft, the Commemorative Air Force Museum houses thousands of other items, including artifacts from the air forces of every major belligerent state that fought in World War II. A highlight of the museum is the popular collection of fuselage artwork, believed to be the largest such collection in the world.

---

158. PRESIDIO LA BAHIA

217 Loop 71, Goliad, Texas, 77963

Site Type: Fortified Mission
Conflicts: Texas Revolution
Dates: Originally completed in 1771
Designations: National Register of Historic Places, National Historic Landmark, Texas Historic Landmark
Web: www.presidiolabahia.org (official website)

The Presidio La Bahia, full name Presidio Nuestra Senora de Loreto de la Bahia, is the largest surviving colonial era stone fortress in America outside of the Eastern Seaboard. Built by the Spanish when their empire still extended into Texas, this fortress was involved in both the Mexican War of Independence against Spain and the Texas War of

Independence against Mexico. It was the sight of the Goliad Massacre in 1836, which helped to galvanize the Texas uprising.

Goliad was home to a series of colonial fortresses as far back as 1721, when the French controlled the area. The current fortress was completed by the Spanish in the 1770s. In 1812 and 1813 the fort was bitterly fought over between Spanish troops and Mexican revolutionaries. The Presidio La Bahia became a fortress of Mexico in 1821.

In 1835, in the early stages of the Texas uprising against Mexico, the Presidio was seized by a small force of Texans. Command of the fort fell to James Fannin, who led an ill-fated attempt to relieve the defenders of the Alamo from the Presidio in February of 1836. Most of the garrison was later captured and nearly all were killed by Mexican troops at the Goliad Massacre. The Presidio subsequently became one of the great memorial sites of Texan independence. The Presidio La Bahia was listed on the National Register of Historic Places and named a National Historic Landmark in 1967 and a Texas Historic Landmark in 1969.

The Presidio La Bahia is actually a fortified mission. The chapel is constructed with especially thick walls and only a handful of apertures. An imposing wall encloses a large courtyard. Although the chapel was largely completed by 1771 much of the current structure dates from restorations in the 1960s.

---

159. PALO ALTO BATTLEFIELD NATIONAL HISTORICAL PARK

7200 Paredes Line Road, Brownsville, Texas, 78526

Site Type: Battlefield - Battles of Palo Alto & Resaca de la Palma
Conflicts: Mexican-American War
Dates: Battles fought on May 8-9, 1846
Designations: National Register of Historic Places, National Historic Landmark, National Historic Park
Web: www.nps.gov/paal (official website)

The Battles of Palo Alto and Resaca de la Palma were the first major engagements of the Mexican-American War. This two-part battle took place in the disputed territory along the border between Texas and Mexico following America's annexation of the former. Despite a significant advantage in available forces, the two battles were humiliating defeats for Mexico and opened the way for an American invasion of the Mexican heartland.

In the early months of 1846 the American annexation of Texas, combined with failed diplomatic efforts between the United States and Mexico, led inexorably to a border clash between the two North American giants. Both sides sent armies to the border, the Americans being led by future president Zachary Taylor. The first clash came at the Battle of Palo Alto on May 8 near Fort Texas, with the Mexicans attempting to besiege the fort and an outside American force attempting to stop them. Despite significantly outnumbering the Americans the Mexicans were driven off with moderate casualties.

The next day an even larger Mexican force was mobilized using additional men from the siege. The two sides met a few miles away at the Battle of Resaca de la Palma. The fighting at this larger battle was fiercer. Despite having even a greater advantage in numbers, the Mexican army was once again forced to retreat, this time back across the Rio Grande River. Both battles were hailed as American victories and helped to rally the American people to support the war. Palo Alto Battlefield was designated a National Historic Landmark in 1960, a National Historic Site in 1978 and a National Historic Park in 2009.

The Palo Alto Battlefield National Historic Park incorporates parts of both battlefields in two separate locations. The main location is Palo Alto, where the first day of fighting took places. A stone monolith marks and commemorates the location of the engagement. A number of artillery pieces of the period can be found scattered around the site.

160. THE ALAMO MISSION

300 Alamo Plaza, San Antonio, Texas, 78205

Site Type: Fortified Mission
Conflict: Texas Revolution
Date: Originally completed in 1758
Designations: National Register of Historic Places, National Historic Landmark, Texas State Antiquities Landmark, UNESCO World Heritage Site
Web: www.thealamo.org (official website)

The Alamo, America's most famous fort, was not in fact built as a fort, but rather as a Catholic mission which was later occupied by Mexico for use as a garrison outpost. Although fortifications and barracks were added in the 18th century the mission was never meant for heavy military duty. Ironically, the Alamo went on to endure the most famous siege in American history. Here in 1836, fewer than two hundred irregular troops and volunteers managed to hold out against thousands of professional Mexican soldiers for the better part of two weeks.

In 1835 the province of Texas broke away from Mexico, kicking off the Texas Revolution. The next year Mexican president and general Antonio Lopez de Santa Ana personally led an army into Texas in order to crush the revolt. The Mexicans drove the defenders back to San Antonio where a garrison of rebels, aided by volunteers from as far away as Tennessee, held their ground. Among the defenders were William Travis, Jim Bowie and former Congressman Davy Crockett.

For thirteen days the tiny force of Texans held out against overwhelming odds. The Alamo finally fell to a massive attack on March 6. However, the action successfully delayed the Mexican army long enough to allow Texas to regroup at the Battle of San Jacinto a little over a month later. The Alamo was retained for military use for the next forty years and then turned over to the state of Texas in 1876. It is now overseen by the Daughters of the Republic of Texas. The Alamo was designated a National Historic Landmark in 1960, added the National Register of Historic Places in 1966, and named a Texas State Antiquities Landmark in 1983. It was included the San Antonio Missions UNESCO World Heritage Site in 2015.

The Alamo Mission was originally constructed in the early 18th century. Fortifications began to be added in the 1750s and were constructed adjacent to the existing stone chapel. After years of neglect the Alamo grounds were systematically renovated in the 20th century. The site is now largely restored to its state at the time of the battle. Although perhaps less visually impressive than other American forts, it is nevertheless one of the most visited, and arguably the most famous, fortress in the United States.

---

161. MILITARY WORKING DOG TEAMS NATIONAL MONUMENT

2434 Larson Street, San Antonio, Texas, 78236

Site Type: Monument – Military Dogs
Dates: Dedicated in 2013
Web: https://myairmanmuseum.org/military-working-dogs (official website)

Military working dogs have been used by the armed forces of the United States since the 19th century. The use of dogs in combat became an especially important part of battlefield operations in jungle warfare against the Japanese Empire and later in Vietnam. Their contribution in these and other conflicts have already been recognized for decades.

In 2008 Congress authorized the establishment of a national monument honoring the work of military dogs and their handlers. A foundation was established and a site was chosen at Lackland Air Force Base due to its historic importance as a training center for dog teams. The Military Working Dog Teams National Monument was completed in 2013.

The Military Working Dog Teams National Monument consists of a wide marble plaza adorned with bronze statues of a German

Shepherd, Belgian Malinois, Labrador Retriever and Doberman Pinscher, all breeds commonly used by the military. There is a particularly heartwarming statue of a soldier pouring water into a helmet for his dog. A marble monument commemorates all those dogs who have died while in military service.

---

162. NATIONAL MUSEUM OF THE PACIFIC WAR

311 East Austin Street, Fredericksburg, Texas, 78624

Site Type: Museum
Conflict: World War II
Dates: Opened in 1969
Web: www.pacificwarmuseum.org (official website)

The National Museum of the Pacific War is a collection of sites related to World War II located in the home town of Admiral Chester Nimitz. Begun as the Admiral Nimitz Museum, the National Museum of the War in the Pacific was added later, along with monuments and a garden. Collectively the entire complex commemorates the war against the Japanese Empire.

Chester Nimitz was born in 1885 in Fredericksburg, Texas where his grandfather owned a hotel. He attended the United States Naval Academy and served in World War I. Nimitz rose through the ranks, and when the United States was drawn into World War II he was named commander of the Pacific Fleet. He ultimately achieved the rank of Fleet Admiral and represented the United States at Japan's surrender in 1945. He died a national hero in 1965.

The National Museum of the Pacific War opened in 1969 in Nimitz's grandfather's hotel. It currently houses a gallery with exhibits on Nimitz's life. The main museum, opened next door in the 1990s, includes exhibits on the history of the War in the Pacific and a large archival collection. A monument out front honors ten presidents who served in various capacities during World War II. An area designated as the Pacific Combat Zone offers a recreation of a battle in the South Seas.

163. FORT DAVIS NATIONAL HISTORIC SITE

101 Lt. Flipper Drive, Fort Davis, Texas, 79734

Site Type: Military Base
Conflict: Later American Indian Wars
Dates: Established in 1854
Designations: National Register of Historic Places, National Historic Landmark, National Historic Site
Web: www.nps.gov/foda (official website)

Fort Davis is one of the finest surviving frontier posts of the United States Army in the American southwest. Constructed primarily to protect travelers in western Texas, its garrison only ever saw light action against raiding Apaches. The fort was annexed by the Confederacy during the Civil War for a brief period and later used as a base for the Buffalo Soldiers. A significant number of the fort's buildings have survived and have been restored.

A few years after the Mexican-American War, settler expansion across the southwestern United States was well underway. Fort Davis was one of the major forts established in the 1850s to protect merchants and settlers as they travelled west towards New Mexico, Arizona and California. Due to its remote location the federal government abandoned the fort in 1861 after hostilities with the South broke out.

The Confederacy took over the fort for about a year, fighting several small skirmishes with local Apaches during their tenure before abandoning the fort once more in 1862. The fort was reoccupied and rebuilt by the army in 1867. Fort Davis was named a National Historic Landmark in 1960, a National Historic Site in 1961 and listed on the National Register of Historic Places in 1966.

Fort Davis is a full-fledged 19th century non-walled military base. The site consists of two dozen surviving buildings surrounding a large, open parade ground suitable for a significant cavalry presence. Several of the buildings on site have been renovated and are now open to the public. There are also remains of many more buildings, which help to give visitors an idea of how large the fort was at its height in the 1880s.

164. WHITE SANDS MISSILE RANGE MUSEUM

WSMR P Route 1, White Sands Missile Range, New Mexico, 88002

Site Type: Military Base, Museum
Dates: Base established in 1945
Web: www.wsmr-history.org (official website)

The White Sands Missile Range Museum is dedicated to the history and testing of missiles used by the American military. Run by the White Sands Missile Range Foundation, it is located on a military facility where missile technology has been developed and tested since the 1940s. The White Sands Missile Range is perhaps most famous for being the location of the test detonation of the first atomic bomb in 1945.

The area in and around White Sands in New Mexico had been used for testing weapons since the early days of World War II. After the surrender of Nazi Germany, captured V-2 rockets were brought here and studied, with many being launched in some of America's earliest missile tests. On July 16, 1945, under the operational code name Trinity, an atomic bomb, the world's first, was detonated here. Less than three weeks later a similar bomb was dropped on Hiroshima in Japan.

The White Sands Missile Range Museum is home to a variety of exhibits, primarily on missiles and nuclear weapons. Over sixty missiles and related pieces of military hardware spanning more than seven decades are on display in the desert outside of the museum. In addition to the military exhibits there are also displays of earlier artifacts from White Sands' days when Native Americans and ranchers lived in the area.

165. SAND CREEK MASSACRE NATIONAL HISTORIC SITE

1301 Maine Street, Eads, Colorado, 81036

Site Type: Battlefield – Battle of Sand Creek
Conflict: Later American Indian Wars
Dates: Massacre took place on November 29, 1864
Designations: National Historic Site
Web: www.nps.gov/sand (official website)

The Battle of Sand Creek was a battle in name only. There was no fighting to speak of, as there were few, if any, Native American warriors present at the time seven hundred federal soldiers attacked and wiped out an encampment of Cheyenne and Arapaho. Although the Sand Creek Massacre was just one of many such incidents along the frontier in the 19th century, there is little doubt that this was among the greatest of the atrocities and one of the darkest stains on the record of the United States Army.

During the Civil War federal troops were dispatched to Colorado to secure the territory and prevent it from falling into Confederate hands. John Chivington, one of the Union commanders stationed in Colorado, had an undying hatred of Native Americans, and took advantage of his authority to harass and persecute the local tribes. By the summer of 1864 tensions were on the rise.

Native Americans deemed friendly were offered the protection of the government, and part of the tribe of a local chief named Black Kettle accepted the offer. Most of the tribe's warriors did not and departed the encampment, leaving behind almost entirely non-combatants. These flew both the American flag and a white flag demonstrating their non-hostility, but this did not avail them on November 29 when Chivington's men rode in and killed or mutilated almost everyone present. The Sand Creek Massacre National Historic Site was established in 2007.

The Sand Creek Massacre National Historic Site is one of the saddest in the National Park system. It incorporates the site of the encampment and a large area beyond. The monuments are minimal, and almost exclusively commemorate the victims of the massacre, though one stone monument on site still refers to the massacre as a battle.

---

166. BENT'S OLD FORT NATIONAL HISTORIC SITE

35110 State Highway 194 East, La Junta, Colorado, 81050

Site Type: Adobe Fort
Conflict: Mexican-American War
Dates: Originally completed in 1833
Designations: National Register of Historic Places, National Historic Landmark, National Historic Site
Web: www.nps.gov/beol (official website)

Bent's Old Fort, nicknamed the Castle of the Plains, was one of the most important trading posts west of Leavenworth prior to the Civil War. Constructed privately by merchants, Fort Bent was used almost exclusively as a trading post and was not actively involved in the Indian Wars. It did, however, serve as a military base during the Mexican-American War.

Westward settlement was well underway by the 1830s. Trade and travel along the Santa Fe Trail between the Americans, Mexicans and Native Americans was beginning to grow significantly, providing opportunities for such entrepreneurs as William Bent and Ceran St. Vrain. Bent and St. Vrain built a trading empire in this virgin territory with an impressive fort at its epicenter. This was one of the largest privately built fortifications in American history.

Fort Bend became a critical transportation hub in the American west. It was used as a hunting base for Kit Carson, and briefly saw military activity during the Mexican-American War. It remained in use from 1833 through 1849 when it was abandoned due to cholera. The fort was largely destroyed in 1852 and replaced with a new fort a few miles away the next year. Bent's Old Fort was designated a National Historic Landmark and National Historic Site in 1960 and listed on the National Register of Historic Places in 1966.

Bent's Old Fort National Historic Site is home to a faithful recreation of the original fort close to its original location. Completed in 1976, the fort is a mostly adobe structure with substantial walls lined with buildings surrounding a large open courtyard. Among the recreated buildings are various guard towers and guest rooms where merchants and travelers would have stayed.

---

167. UNITED STATES AIR FORCE ACADEMY

2304 Cadet Drive, Air Force Academy, Colorado, 80840

Site Type: Military Academy
Dates: Opened in 1954
Web: www.usafa.af.mil (official website)

The United States Air Force Academy is the primary college for the training of American Air Force officers. Established in 1954, it was the last of America's major military academies to open. Although most of the campus is not accessible to visitors, especially when school is in session, there is a visitor center, and limited tours of the academy are sometimes available.

The Air Force spent the better part of the first half of the 20th century as an extension of the army. It did not become an independent branch of the military until 1947. Training of air officers was not prioritized during this early period. However, with the formation of an independent Air Force, the call for an independent academy gained traction, and in 1954 the USAFA was finally established.

The first Air Force cadets began their education at a temporary site in Denver before moving to Colorado Springs in 1958. Many of the academy's graduates served in the Air Force throughout the Cold War, notably in Vietnam. Many astronauts from the United States Space Program attended the Air Force Academy.

Tours of the United States Air Force Academy begin at the Barry Goldwater Visitor's Center, which has exhibits on the academy and cadet life. The academy's most distinctive building, and a highlight of any visit, is the spectacular cadet's chapel. Arguably the most futuristic of any collegiate chapel in the United States, this masterpiece of modern architecture is evocative of a rocket preparing for liftoff.

---

168. CHEYENNE MOUNTAIN COMPLEX

1, 101 Norad Road, Colorado Springs, Colorado, 80906

Site Type: Military Command Center
Conflict: Cold War
Dates: Opened in 1966
Web: www.norad.mil/about-norad/cheyenne-mountain-air-force-station (official website)

The Cheyenne Mountain Complex is one of America's best known Cold War military installations. Made famous by the 1983 movie *Wargames*, the Cheyenne Mountain Complex is a massive subterranean command bunker designed to withstand a nuclear strike. It is located close to Peterson Air Force Base and is commonly associated with the North American Aerospace Defense Command (NORAD).

Throughout the 1950s the threat of an all-out nuclear war between the United States and the Soviet Union was very real, and necessitated new military installations capable of surviving such an attack. Between 1961 and 1966, the most famous of these was built in the mountains outside of Colorado Springs. This massive bunker complex would go on to host the Space Defense Center as well as the National Civil Defense Warning Center and other military commands.

Although many believe that NORAD is based here, the Cheyenne complex is actually an alternate NORAD base, presumably to be activated in the event of nuclear war. Numerous apocalyptic stories have been set here, from *Dr. Strangelove* to *The Terminator*. The Cheyenne Mountain Complex is also home to a popular American Christmas tradition, the NORAD Santa Tracker.

While the Cheyenne Mountain Complex is not top secret, it is one of the most heavily fortified military sites in the United States. In fact one of the most popular things to see at the complex are the twenty-five ton blast doors which protect the main bunker and control center. Tours of the facility are possible but must be arranged in advance. Access to visitors at any given time is likely to be extremely limited.

---

169. COVE FORT HISTORIC SITE

Highway 161se, Beaver, Utah, 84713

Site Type: Masonry Fort
Dates: Originally completed in 1867
Web: www.covefort.com (official website)

Cove Fort is one of the better preserved frontier forts in the far American west. This is due in large part to its stone construction, rather than wood or adobe. The fort was built by Mormons for use as a way station and remained an important transportation hub throughout the second half of the 19th century.

By the 1860s Mormon settlers had become well established in Utah, with new towns under construction throughout the territory. Around this time Brigham Young, the leader of the community, gave thought to protecting the settlements and travelers. One of his projects was the construction of a fort along the Mormon trail between the settlements at Filmore and Beaver.

Cove Fort was completed a few years later, becoming one of the most important and busiest transit points in the entire Utah Territory. It remained in active use until the 1890s when it was sold off to private ownership. The fort was acquired by the Church of Latter Day Saints in 1989, restored and reopened as an historic site.

Cove Fort is a very well preserved fortification thanks primarily to its stone construction, a rarity in the early far west territories. Fully restored in the 1990s, it looks today much as it did a century ago. Several of the fort's buildings can be visited, including the common kitchen and washing areas, the telegraph office and the rooms where Ira Hinckley, who oversaw the original construction of the fort, lived with his family.

---

170. FORT BOWIE NATIONAL HISTORIC SITE

3500 South Apache Pass Road, Bowie, Arizona, 85605

Site Type: Military Base
Conflict: Later American Indian Wars
Dates: Established in 1864
Designations: National Historic Site
Web: www.nps.gov/fobo (official website)

Fort Bowie was one of the most remote fortifications in America at the time of its construction in the mid-19th century. It was built to protect access through the Apache Pass following several engagements between federal troops and the local tribes. It served as a major base of operations in the area for the United States Army following the Civil War.

During the early 1860s clashes between the federal government and the Apaches in Arizona became increasingly frequent, a problem that exasperated the Union war effort against the Confederacy in that area. In 1861 Apache raids on settlements led to the capture of the famous warrior Cochise, who subsequently escaped. In July of the next year a major Apache force led by Cochise attacked a column of federal soldiers at the Battle of Apache Pass.

Although the army drove the Apaches off, the event had disrupted a minor campaign against the Confederacy in New Mexico. To discourage further attacks the army built Fort Bowie. The fort became the primary base for military operations against the Apaches in the years after the Civil War. Geronimo was briefly incarcerated here following his capture in 1886. The site remained in operation until 1894, when it was abandoned. Fort Bowie was designated a National Historic Site in 1960.

Fort Bowie National Historic Site, while one of the more historic frontier forts in the far Southwest, is not particularly well preserved and in fact consists mostly of ruins, albeit in a spectacular setting. The surviving sections, mostly building foundations and lower walls, give a rough idea of the layout and scope of the base. Parts of the cemetery have been restored in recent years.

---

171. DAVIS-MONTHAN AIR FORCE BASE AIRCRAFT BONEYARD

South Wilmot Road, Tucson, Arizona, 85708

Site Type: Military Base
Dates: Opened in 1946
Web: www.military.com/base-guide/davis-monthan-air-force-base (official website)

The Davis-Monthan Air Force Base Aircraft Boneyard is the largest storage facility for outdated military planes in the United States and one of the largest in the world. Officially known as the 309th Aerospace Maintenance and Regeneration Group, the Boneyard is currently home to over four thousand aircraft. While some are kept intact, most are destined to be sold or turned into scrap.

The first decommissioned aircraft to arrive at Davis-Monthan were moved here in 1946 and consisted mostly of World War II bombers. Over the next few decades more and more outdated planes were retired here. The Boneyard became even more crowded as other storage facilities were closed and their inventories consolidated here in the 1960s.

The Davis-Monthan Boneyard is home to a huge collection of aircraft that dates back over seventy years. There are fighters, bombers, helicopters and all manner of support aircraft representing all branches of the United States military. There are even decommissioned missiles. Davis-Monthan Air Force Base is an active military installation, and visits to the Boneyard are by tour only. Access may be subject to restrictions for visitors.

172. FORT CHURCHILL STATE HISTORIC PARK

10000 Highway 95A, Silver Springs, Nevada, 89429

Site Type: Military Base
Conflict: Later American Indian Wars
Dates: Established in 1861
Designations: National Register of Historic Places, National Historic Landmark, Nevada State Park
Web: http://parks.nv.gov/parks/fort-churchill (official website)

Fort Churchill was constructed to protect the Central Overland Route from Salt Lake City to Carson City. A major way station for travelers heading towards California, it was necessitated by the increasing friction between the army and local Native American tribes, mainly caused by the army. It also served as a supply depot for the Union during the Civil War.

Fort Churchill had its origins in Williams Station, a Pony Express stop along the Carson River in the 1850s. In 1860 the army arrived in the area and fought a short war with the local Paiute and Bannock Indians. This conflict, known as the Pyramid Lake War, was precipitated by the rape of two girls from the Paiute tribe and the subsequent raids on settlers. This conflict saw one of the few major victories of Native Americans over the United States Army at the First Battle of Pyramid Lake.

In the end the tribes were defeated, and Fort Churchill constructed in 1861 as a deterrent to further unrest. The fort helped to protect the supply routes to Northern California during and immediately after the Civil War, but within a decade had become obsolete as an outpost and was abandoned in 1869. Over the next century the site was held privately for a time before being turned over to the state. Fort Churchill was named a National Historic Landmark in 1961, listed on the National Register of Historic Places in 1966 and made a Nevada State Park in 1957.

Fort Churchill was largely in a state of disrepair when restorations began in the 1930s. Although most of the fort is ruins, a few of the original buildings are intact. This includes the Buckland Station, a building used by the Pony Express, located outside of the fort.

NORTHWEST AND ALASKA

173. FORT LARAMIE NATIONAL HISTORIC SITE

965 Grey Rocks Road, Fort Laramie, Wyoming, 82212

Site Type: Military Base
Conflict: Later American Indian Wars
Dates: Established in 1834
Designations: National Register of Historic Places, National Historic District
Web: www.nps.gov/fola (official website)

Fort Laramie was one of the longest active forts on the northwestern frontier, and was in use for a period of sixty years. It began as a trading post in the 1830s, saw action before, during and after the Civil War, and only became obsolete after the completion of the Transcontinental Railroad in the 1890s. It was a critical stop along the Oregon Trail and an important base of military operations during the late 19th century.

The area near Laramie saw the arrival of white settlers, mostly trappers, as early as the 1810s. Within twenty years it had become a popular stopping point for settlers heading towards Utah, Oregon and California, and in 1834 Fort Laramie was built by private interests to protect the route. The fort was acquired by the United States Army in 1849. Soldiers from the fort were involved in some of the earliest clashes with the tribes of the Great Plains, including fighting with the Sioux that left nearly thirty soldiers dead at Grattan.

After the Civil War the United States government came into increasing conflict with the Plains Indians, and Fort Laramie became a critical base of operations for the army. At the same time the fort became less important economically as traffic on the Oregon Trail started to wane. In 1890 the fort was abandoned and sold off to homesteaders. Fort Laramie was listed on the National Register of Historic Places in 1966 and named a National Historic District in 1983.

Fort Laramie National Historic Site is an expansive park that includes what survives of the 19th century frontier fort. There are about a dozen buildings still standing around the perimeter of the parade ground, including some of the barracks and the commissary. There are also ruins of other original buildings scattered around the base.

---

174. LITTLE BIGHORN BATTLEFIELD NATIONAL MONUMENT

756 Battlefield Tour Road, Crow Agency, Montana, 59022

Site Type: Battlefield – Battle of Little Bighorn
Conflict: Later American Indian Wars (Sioux Wars)
Dates: Battle fought on June 25-26, 1876
Designations: National Monument
Web: www.nps.gov/libi (official website)

The Battle of Little Bighorn was arguably the most famous engagement in the centuries-long series of wars between the United States Army and the continent's Native American tribes. Featuring two of the America's most famous antagonists, George Armstrong Custer and Crazy Horse, it was also the greatest Native American victory in history. Unfortunately it failed to put a stop to the relentless drive of pioneers and settlers flocking to the American west. The Battle of Little Bighorn made Crazy Horse a household name, and in the end he became a hero not only among the various western tribes but to many Americans as well. The battle is now commemorated by the descendents of both sides.

In the years following the Civil War American settlers started heading westward in huge numbers. This was exacerbated by discovery of gold in the Black Hills, putting great pressure on the tribes that had lived there for centuries. From 1868 to 1875 the two sides managed to maintain a semblance of peace; but in 1876 open hostilities broke out in the Great Sioux War. Thousands of Native Americans arose in defiance of the broken treaties and set out to protect their land. In response the United States began sending in large military forces to deal with the rebellion, including the Seventh Cavalry under the command of Custer.

In June Custer's cavalry set out to crush the locals. However, he seriously underestimated his opponents and foolishly divided his forces. Thanks to this tactical blunder, and to superior scouting on the part of Crazy Horse's army, Custer's troops found themselves outnumbered at every turn. Eventually cornered on what is now known as Custer Hill, a large part of the Seventh Cavalry was wiped out at Custer's Last Stand. This victory was the greatest moment of glory for the Native Americans in the Indian Wars, but in the end only put off the final day of reckoning for a short time.

Little Bighorn Battlefield National Monument consists of several sites along the Little Bighorn River that are associated with the battle. These include the large exposed field where Custer's Last Stand took place and the cemetery where most of the fallen from the battle are buried. A large obelisk-like monument stands in the cemetery to commemorate the engagement and the fallen.

---

175. BIG HOLE NATIONAL BATTLEFIELD

16425 Highway 43 West, Wisdom, Montana, 59761

Site Type: Battlefield – Battle of the Big Hole
Conflict: Later American Indian Wars
Dates: Battle fought on August 9-10, 1877
Designations: National Battlefield, National Historic Park
Web: www.nps.gov/biho (official website)

The Battle of the Big Hole was one of the most famous engagements in the later years of the Indian Wars. Fought between the army and the Nez Perce tribe, the battle did not end in a clear-cut victory for the United States, an extreme rarity at the time. The battle was a successful delaying action by the Nez Perce, who were attempting to flee American jurisdiction and escape to Canada.

The Nez Perce War of 1877 was part of the massive effort of the United States Army to bring all of the Plains tribes under control following the disastrous defeat of the Seventh Cavalry at Little Bighorn the year before. The Nez Perce, under several leaders which included Chief Joseph, decided the best course of action was to relocate to Canada and join other tribes that had taken refuge there.

The tribe's trek to Canada was dogged the entire way by the army, and the Nez Perce were forced to fight several delaying actions, at which they were reasonably successful. In August a cavalry force under John Gibbon caught up with the Nez Perce at Big Hole. Fighting between the reasonably evenly matched forces was brutal, and both sides took heavy casualties. While the Americans held the field afterwards, they were prevented from pursuit by skilled Nez Perce snipers who served as the rearguard. The Big Hole National Battlefield was established in 1963 and incorporated into the Nez Perce National Historic Park in 1965.

Big Hole National Battlefield is one of the most beautiful battlefield parks in the United States. Its magnificent setting of grasslands and hills, with snow covered peaks in the distance, belies the fighting that took place here in 1877. There are markers on the site noting where points of the battle took place, as well as monuments to both sides. Both teepees and cannons are on display near the visitor center.

---

176. FORT HALL

3001 Avenue of the Chiefs, Pocatello, Idaho, 83204

Site Type: Masonry Fort
Dates: Originally completed in 1834
Designations: National Register of Historic Places,
National Historic Landmark
Web: www.forthall.net (official website)

Fort Hall was one of the most isolated mercantile posts in the Pacific Northwest in the late 19th century. Originally established in the 1830s to support the fur trade, it later became a critical stop along the Oregon Trail. It was later replaced by a new Fort Hall. Both forts were eventually turned over to local Native American tribes for their use.

The first Fort Hall was built in 1834 essentially as a supply depot for early settlers in the Oregon Territory. The Oregon Trail, which became a popular route for West Coast bound settlers in the 1840s, passed the fort just before it split off from the California Trail, adding to Hall's importance.

The original fort was abandoned during the Civil War, and a new Fort Hall constructed in 1870 about twenty five miles away. This was used by the United States Army until 1883, when it was turned over to the Bureau of Indian Affairs for tribal use. Old Fort Hall was named a National Historic Landmark in 1961 and listed on the National Register of Historic Places in 1966.

The modern Fort Hall is actually a faithful replica of the original structure. There is nothing left of the old fort, although the site is marked with a memorial plaque. The replica is dominated by an impressive watch tower. The site features recreations of various log cabin facilities as well as a large teepee in the center of the compound.

---

177. FORT CLATSOP

92343 Fort Clatsop Road, Astoria, Oregon, 97103

Site Type: Wooden Stockade Fort
Dates: Originally completed in 1805
Designations: National Register of Historic Places
Web: www.nps.gov/lewi (official website)

Fort Clatsop was one of the first outposts ever constructed by agents of the United States government in the Pacific Northwest. It was built under the auspices of Lewis and Clark during their famous expedition to explore the newly acquired territory on the American frontier. Only used for a few months, the fort was later turned over to local Native American tribes and used as a fur trading post.

The Lewis and Clark Expedition made its way into what is now Oregon in 1805 after a year-long trip along the rivers of the Northwest. Upon arriving on the Pacific Coast they constructed Fort Clatsop, where they spent the winter of 1805-1806. In March of 1806 the expedition turned the fort over to the local Clatsop Indians and departed.

The original fort only survived for a few decades, but during this time it served as an important fur trading post. Largely abandoned due to disrepair by the mid-19th century, the fort was reconstructed in the 1950s, and again in 2006. Fort Clatsop was listed on the National Register of Historic Places in 1966. It is also part of the multi-site Lewis and Clark National Historic Park.

Fort Clatsop is entirely a replica based on descriptions of the original fort, which eventually crumbled under the onslaught of poor weather conditions. Another prior replica constructed in 1955 was destroyed by fire in 2005. The current incarnation consists of several log buildings and a stockade, and stands close to what is believed to be the site of the original structure.

---

178. FORT YAMHILL BLOCKHOUSE

402 3rd Street, Dayton, Oregon, 97114

Site Type: Wooden Blockhouse
Dates: Originally completed in 1856
Web: www.daytonoregon.org/explore-dayton
(official city tourism website)

Fort Yamhill was one of the more substantial early forts located in the Oregon Territory, even though it was only active for about a decade. It was noteworthy for the famous officers that served there. Its blockhouse is one of the best surviving such structures in the Pacific Northwest. The blockhouse was moved to Dayton in the early 20th century, while the rest of the fort is forty miles to the west in Grand Ronde.

Settlement in the Oregon Territory was on the rise following the California Gold Rush, and by 1856 the United States Army had constructed Fort Yamhill to protect pioneers and fur traders in the area. Among the officers who served at the fort were Joseph Hooker, Joseph Wheeler and Philip Sheridan, all of whom went on to serve as generals during the American Civil War.

The fort was decommissioned in 1866, barely a year after the war ended. The blockhouse later served as a prison, and was ultimately moved to nearby Dayton in 1911. As of the time of this writing there were plans to reconstruct the fort on its original site. The blockhouse of Fort Yamhill was listed on the national Register of Historic Places in 1971.

The Fort Yamhill Blockhouse is a well preserved 19th century wooden structure overlooking Dayton's courthouse square. Standing two stories in height, the blockhouse is covered in rifle holes facing out in all directions. At the time of this writing it was unknown if the blockhouse interior was accessible. There is little left at the Fort Yamhill site in Grand Ronde, although the small house which served as the officer's quarters is still standing.

---

179. FORT VANCOUVER NATIONAL HISTORIC SITE

1501 East Evergreen Boulevard, Vancouver, Washington, 98661

Site Type: Wooden Stockade Fort
Dates: Originally completed in 1824
Designations: National Register of Historic Places, National Historic Site
Web: www.nps.gov/fova (official website)

Fort Vancouver was one of the first significant fortifications to be established on the West Coast. Originally built by the Hudson Bay

Company as their Pacific headquarters, it was among the most important trading posts along the West Coast prior to the California Gold Rush. It was taken over by the United States Army a few years after the Treaty of Oregon was signed recognizing the fort as being in American territory.

Fur hunters and trappers were active in the Oregon Territory as far back as the 18th century. In 1824 the Hudson Bay Company established a permanent, fortified fur trading post in what is now Vancouver, Washington. Traders from the post were active from Alaska to California. The fort remained in use by the Hudson Bay Company until 1860 when a new headquarters was established in Canada.

After their departure the fort was taken over by the United States government, as the army had already established a barracks nearby. The original fort was destroyed in a fire in 1866, but an active military base was maintained on the site. One of the first military airfields on the West Coast was established here in 1911. Fort Vancouver was designated a National Historic Site in 1961.

The Fort Vancouver National Historic Site is one of the oldest surviving military installations on the West Coast. Though there is nothing left of the original fort, part of the wooden stockade has been recreated, and the 1849 barracks are still there. Pearson Field, which served as an Army aviation base from 1923-1941, is located here and is now home to the Pearson Air Museum.

---

180. ALEUTIAN WORLD WAR II NATIONAL HISTORIC AREA

2716 Airport Beach Road, Unalaska, Alaska, 99865

Site Type: Battlefield – Battle of Attu
Conflict: World War II
Dates: Battle fought on May 11-30, 1943
Designations: National Register of Historic Places, National Monument, Alaska Heritage Site
Web: www.nps.gov/aleu (official website)

The Aleutian World War II National Historic Area commemorates the American defense of the Aleutian Islands against the Japanese in the early stages of the war in the Pacific. Although the engagement is sometimes known as the Battle of Attu, fighting took place across numerous islands, including nearby Dutch Harbor. The national historic area incorporates several distinct places of interest on different islands, including the B-24D Liberator Crash Site on Atka.

In June of 1942 a Japanese force seized several islands in the Aleutian chain off the coast of Alaska. Whether this was a forerunner to an invasion of Alaska or merely a diversionary tactic for the Japanese fleet heading towards Midway, the Americans took it seriously. A force was dispatched to drive the Japanese out, a campaign which took over a year. The last Japanese occupiers withdrew from the Aleutians in the summer of 1943.

In December of 1942 a B-24D Liberator was forced to crash land on the island of Atka due to bad weather. Although there was only one casualty, the plane was effectively destroyed. The crew was rescued the next day and the Liberator abandoned. Amazingly the wreckage was left undisturbed, and has mostly survived to the present day. The Atka B-24D Liberator Crash Site was listed on the National Register of Historic Places in 1979.

The Aleutian World War II National Monument consists of three separate areas on different islands. The main site of interest is Fort Schwatka, the highest coastal battery ever built on American soil. The B-24D Liberator wreck, located on Atka Island, is one of the last surviving remnants of the American campaign to defend the Aleutians from the Japanese. The wreck includes more than half of the original plane, the remains of which are in surprisingly good condition. Most of the forward fuselage is intact, as are most of both wings and the engines.

CALIFORNIA & PACIFIC ISLANDS

181. ALCATRAZ ISLAND

Alcatraz Island, San Francisco, California, 94133

Site Type: Masonry Fort
Dates: Originally completed in 1859
Designations: National Register of Historic Places, National Landmark District
Web: www.nps.gov/alca (official website)

Alcatraz Island was the site of America's most infamous prison from 1934 until 1963. It is for this reason that the small island in San Francisco Bay is so well known. But the island's history goes back almost a century further, when it was home to first a military fortress and then a military prison.

The potential of Alcatraz Island as a military outpost was realized as early as the 18th century when the bay area was scouted by the Spanish. The island was acquired by the United States in 1848, and a fortress completed a few years later. By the time the Civil War started Alcatraz was the most heavily fortified structure on the West Coast.

By the late 1860s the increasingly obsolete fortress was being used to house military prisoners, and in 1868 was converted outright to a detention facility. It became the Western Military Prison in 1907, which it remained until 1933. Over the course of seven decades the prison housed POWs from the Confederacy and Spain, as well as American conscientious objectors. Alcatraz became a non-military penitentiary in 1934. Alcatraz Island was listed on the National Register of Historic Places in 1976 and named a National Historic Landmark District in 1986.

Alcatraz Fortress was largely incorporated into the prisons that were later built on the island, and some of these areas can be seen on tour. Highlights of the military era include the chapel, completed in the 1920s, the original military parade grounds, and the remains of the officer's club which was destroyed in a fire in 1970.

---

182. FORT POINT NATIONAL HISTORIC SITE

Long Avenue & Marine Drive, San Francisco, California, 94129

Site Type: Masonry Fort
Dates: Originally completed in 1861
Designations: National Historic Landmark, California Landmark
Web: www.nps.gov/fopo (official website)

Fort Point is one of the oldest American military installations on the West Coast. It was used on and off for about a century, protecting San Francisco Harbor from enemy naval threats during the American Civil War and both World Wars. Thanks to its location beneath the Golden Gate Bridge, it is possible to look straight down at the fort without having to be in a plane or helicopter.

The Spanish constructed an early fort on the current site in 1794, more or less marking the northern end of their empire on the West Coast of the Americas. It changed hands to Mexico in 1821 before being largely abandoned. In 1846 what was left of the site was occupied by the United States. By the outset of the Civil War a new, modern fort was completed in order to defend the bay.

Over the next few decades Fort Point saw a variety of uses, including as a military prison. The fort was brought into service again during World War II to support the harbor defenses against a possible Japanese submarine attack. Fort Point was listed on the National Register of Historic Places and named a National Historic Site in 1970.

Fort Point is located at the entrance to San Francisco Bay, directly under the southern end of the Golden Gate Bridge. Because of this it is probably one of the most photographed forts in the United States. Much of Fort Point was restored in the years after World War II. Some of its 19th century artillery pieces are on display in the casements.

183. USS PAMPANITO & SAN FRANCISCO MARITIME NATIONAL HISTORICAL PARK

Pier 45, San Francisco, California, 94133

Site Type: Submarine
Conflict: World War II
Dates: Commissioned in 1943
Designations: National Register of Historic Places, National Historic Landmark
Web: https://maritime.org (official website)

The USS Pampanito is one of the best surviving World War II submarines of the Balao class. Built in less than four months, she was commissioned and active in the Pacific Theater before the end of 1943. The Pampanito patrolled the central part of the Pacific for most of the war.

On September 12, 1944 the Pampanito's patrol sunk the Japanese passenger ship SS Rakuyo Maru, not realizing she was carrying over a thousand Allied prisoners of war. The Pampanito was decommissioned after the war ended, but saw service again in the 1960s as a training vessel. She became a museum ship at Fisherman's Wharf in 1982. The USS Pampanito was named a National Historic Landmark and added to the National Register of Historic Places in 1986.

The USS Pampanito is now part of the San Francisco Maritime Historic Park, located at the city's famous Fisherman's Wharf. In addition to being open for tours, the submarine has facilities to accommodate overnight guests as well as a functional Ham radio room. The maritime park is also home to a number of other historic ships dating from the late 19th and early 20th centuries.

184. WEST COAST MEMORIAL TO THE MISSING

1351 Washington Boulevard, San Francisco, California, 94129

Site Type: Monument – Seamen Killed in World War II
Conflict: World War II
Date: Dedicated in 1965
Website: www.abmc.gov/cemeteries-memorials/americas/west-coast-memorial (official website)

The West Coast Memorial to the Missing commemorates those who gave their lives and whose bodies were never recovered while defending the Pacific Coast during World War II. Over four hundred servicemen from all branches of the armed forces are honored here.

World War II saw many naval engagements across the expanse of the Pacific, with most of the fighting taking place in distant seas. However, Japan did make forays into American coastal waters, usually by submarine, in the early years of the conflict. Not surprisingly these engagements left many sailors and airmen missing in action, lost at sea. The West Coast Memorial to the Missing was erected in 1965 in their honor.

The West Coast Memorial to the Missing of World War II occupies a spectacular setting in San Francisco's Presidio, high above the Pacific Ocean. The large wall-like monument is engraved with the names of those honored. A statue of Athena, shield in hands, stands guard over the monument.

185. ABRAHAM LINCOLN BRIGADE MONUMENT

5 The Embarcadero, San Francisco, California, 94105

Site Type: Monument – Lincoln Brigade
Conflict: Spanish Civil War
Dates: Dedicated in 2008
Website: None Available

The Abraham Lincoln Brigade Monument is one of the most unusual military memorials in the United States, commemorating a little known chapter in American military history. It honors those volunteers who fought and died to prevent the rise of fascism in Spain in the 1930s, a precursor to the fighting that would later ravage the rest of Europe.

From 1936 through 1939 Spain was wracked by a horrific civil war that served as an ominous foreshadowing of World War II. The main factions were backed by Nazi Germany and Soviet Russia. While the Spanish communists were not well liked, the fascists were deemed to be the greater threat and the greater evil. In 1937 volunteers from the United States travelled to Spain to oppose them.

Over the next two years more than three thousand Americans serving in six all-volunteer military units participated in the war. Nearly one in five of these were killed before the Spanish Republic fell and the international brigades were forced to withdraw. Many of those who fought in Spain later went on to fight against the Axis in World War II.

The Abraham Lincoln Brigade Monument in San Francisco was erected in 2008, seven decades after the conflict. The highly unusual steel and glass wall monument houses inscriptions about the conflict and photos of some of those who served. As of the time of this writing the monument was undergoing substantial repairs.

---

186. USS HORNET MUSEUM

707 West Hornet Avenue, Alameda, California, 94501

Site Type: Aircraft Carrier
Conflicts: World War II, Vietnam War, Cold War
Dates: Commissioned in 1943
Designations: National Register of Historic Places, National Historic Landmark, California Historical Landmark
Web: www.uss-hornet.org (official website)

The USS Hornet (CV-12), a surviving sister ship of the Yorktown and Intrepid, has a history very similar to its more famous siblings. It participated in many actions in the Pacific Theater during World War II and Vietnam, and was actively involved in the Apollo space program. Arguably its most famous moment was as the recovery ship for the Apollo craft from the first moon landing. The Hornet also has a reputation for paranormal activity, and has been investigated for such on several occasions.

The Hornet was commissioned in 1943 and went immediately into service in the war against Japan. It was present at the Battle of the Philippine Sea, the Battle of Leyte Gulf, and at the Marianas Turkey Shoot. The Hornet was the second most successful aircraft carrier in the war after the Essex in terms of downed enemy aircraft.

After World War II the Hornet was decommissioned and recommissioned several times, including for service in Vietnam. In 1969 the Hornet recovered Buzz Aldrin, Neil Armstrong and Michael Collins after the first moon landing. She became a museum ship in 1998. The USS Hornet was added to the National Register of Historic Places in and designated a National Landmark in 1991, and designated a California Historical Landmark in 1999.

The USS Hornet Museum is home to a collection of aircraft from World War II and the Cold War. But besides these aircraft and the ship itself, the big draw here are the exhibits from the Apollo Space Program. The exhibit includes an Apollo Command Module used in an unmanned test flight and a mobile quarantine facility. The location where the first steps taken by the Apollo 11 astronauts after their return to Earth is marked on the hangar deck.

---

187. ROSIE THE RIVETER HOME FRONT NATIONAL HISTORICAL PARK

1414 Harbour Way South #3000, Richmond, California, 94804

Site Type: Museum
Conflict: World War II
Dates: Opened in 2000
Designations: National Register of Historic Places, National Historic Park
Website: www.nps.gov/rori (official website)

Officially known as the Rosie the Riveter World War II Home Front National Historic Park, this monument commemorates the millions of workers who supported the war effort, both in Richmond and across the country. As the name implies the park focuses on the role of women in the war effort, but not exclusively. The historic park is relatively new and development is currently ongoing.

The Home Front during World War II was possibly the most energetic time in American history. After years of Depression, everybody was suddenly back to work. With millions of men serving in the military, the factories leaned heavily on female workers to build the guns, tanks, planes and ships needed for victory. Rosie the Riveter became the iconic symbol of women in the workforce.

During this period the city of Richmond was one of the most important production centers on the West Coast. Countless items for the war effort were produced in the factories and shipyards here, including over seven hundred ships in less than four years. Richmond became emblematic of the home front, which is why it was chosen as the site to commemorate the era. The World War II Home Front National Historic Park was established in 2000 and listed on the National Register of Historic Places in 2001.

Rosie the Riveter Home Front National Historical Park is home to a collection of sites that recall the various industries active here in the 1940s. Probably the most interesting destination at the park is the SS Red Oak, a Victory ship actually constructed in Richmond. It is located at the Richmond Shipyards, where hundreds of these ships were built. Also on hand is the Ford Richmond Plant, where over a hundred thousand motor vehicles were assembled prior to being shipped off to war.

---

188. PORT CHICAGO NAVAL MAGAZINE NATIONAL MEMORIAL

4202 Alhambra Avenue, Concord, California, 94553

Site Type: Monument – Port Chicago Disaster
Conflict: World War II
Dates: Dedicated in 1994
Web: www.nps.gov/poch (official website)

The Port Chicago disaster was the most devastating military accident to occur on American soil during World War II. On July 17, 1944 a bomb being loaded onto a cargo ship at the Port Chicago facility accidentally detonated. This set off more bombs that were stored nearby, starting a chain reaction that ultimately led to an immense explosion so powerful that it destroyed much of the surrounding area and killed or wounded more than seven hundred sailors and civilians.

Most of those killed were African American sailors. These sailors had not been properly trained for the handling of munitions, nor were proper safety measures being observed at the site. This incident led to a mutiny of African American sailors who refused to load munitions and who later served several years in military prison. Together the disaster and mutiny were contributing factors that led to the desegregation of the United States Navy.

The Port Chicago Naval Magazine National Memorial commemorates both the military and civilian victims of the accident, as well as those who were court marshaled for their mutiny afterward. It consists of a small plaza by the shore near the site of the incident with several stone monuments. The memorial is located on the grounds of the Concord Naval Weapons Station and can only be visited by tour. As of the time of this writing there were plans to expand the facility.

---

189. TULE LAKE WAR RELOCATION CENTER

800 Main Street, Tule Lake, California, 96134

Site Type: Internment Camp
Conflict: World War II
Dates: Opened in 1942
Designations: National Register of Historic Places, National Historic Landmark, California Landmark
Web: www.nps.gov/tule (official website)

The Tule Lake War Relocation Center is a reminder of one of the sadder chapters of American military history. During World War II, fear of hostile activity by Japanese Americans led the United States government to incarcerate over a hundred thousand people in concentration camps. Citizens and non-citizens alike were stripped of their rights, some until well past the ending of the war. The Tule Lake camp has been preserved in memory of the suffering of Japanese Americans during the war.

Following the Japanese surprise attack against Hawaii in 1941, a panicked American populace turned to their political leaders for increased safety measures. The government responded by turning against residents in the United States who were of Japanese heritage, including tens of thousands who were American citizens. These were rounded up under an executive order issued by President Roosevelt.

Ten concentration camps were established to house those forced to leave their homes. The most infamous of these camps was the one at Tule Lake, where conditions were particularly harsh. The camp here remained open until 1946, well past the Japanese surrender and the end of the war in the Pacific. The legacy of illegal incarceration, lost property and even lost citizenship followed many of those incarcerated to the ends of their lives. The Tule Lake War Relocation Center was listed on the National Register of Historic Places and designated a National Historic Landmark in 2006.

The Tule Lake War Relocation Center is partially preserved, with sections of camp housing still standing behind fences crowned with barbed wire. There are a few surviving common buildings as well as a guard tower. Most of these buildings can be seen by tour only. The visitor's center houses exhibits on the history of the camp.

---

190. UNITED STATES NAVY SEABEE MUSEUM

3201 North Ventura Road, Port Hueneme, California, 93043

Site Type: Museum
Dates: Opened in 1946
Web: www.seabeehf.org/about/museums-heritage-center (official website)

The United States Navy Seabee Museum is the largest museum honoring America's fabled Construction Battalion (CB = "Sea Bee"). There are branch facilities in Gulfport, Mississippi and Daviesville, Rhode Island which are also home to Seabee museums. Located just outside of the Naval Base Ventura County, the museum's collection features exhibits on the battalion's history, mostly from activity in the Pacific region during World War II.

The Seabees were formed in 1941 in response to the need for a mobile force of specialized engineers who could construct support facilities for military use under combat conditions. Trained to build almost anything under tight time constraints, they could also fight when necessary. They became famous for their many heroic efforts in the Pacific Theater against the Japanese during World War II.

The United States Seabee Museum was originally founded in 1946, and the current facility was completed in 2010. Exhibits feature artifacts, including specialized vehicles and equipment, from the thousands of projects the Seabees have completed during their history. Among the collection are many of the cruise books kept by members of the Seabee units.

---

191. USS IOWA MUSEUM SHIP

250 South Harbor Boulevard, Los Angeles, California, 90731

Site Type: Battleship
Conflicts: World War II, Korean War, Cold War
Dates: Commissioned in 1943
Web: www.pacificbattleship.com (official website)

The USS Iowa (BB-61), or the *Big Stick*, was the prototype of the Iowa-class battleships and one of the last battleships ever to be built by the United States. It was one of only a handful of American capital ships to serve in both the Atlantic and Pacific theaters of World War II. The Iowa also played a supporting diplomatic role in Allied relations at the height of the conflict. The USS Iowa was established as a museum ship in 2012.

The Iowa was the first battleship of her class and went into service in early 1943. She spent her first year active in the Atlantic, with two very high profile missions. The first was as a deterrent to the German battleship Tirpitz, which threatened Allied shipping in the North Atlantic. The second was to transport President Roosevelt and other key officials across the Atlantic en-route to the Yalta conference with Churchill and Stalin, where much of the future course of the war was determined.

In early 1944 the Iowa was relocated to the Pacific where she fought in numerous engagements, including the Battle of the Philippine Sea. The Iowa was nearby when Douglas MacArthur famously returned to the Philippines, and was present in Tokyo Bay with the USS Missouri when the Japanese surrendered. She was decommissioned and recommissioned several times, supported military operations in Korea and patrolled the Persian Gulf during the Iran-Iraq War. The USS Iowa was permanently decommissioned in 1990.

The USS Iowa Museum is permanently located at the Los Angeles World Cruise Center, and is one of the only museum battleships not to be located in its namesake state. Among its unique features are its sixteen inch guns, the largest ever placed on an American warship, and the room where President Roosevelt stayed on his voyage across the Atlantic.

---

192. NORTHWOOD GRATITUDE AND HONOR MEMORIAL

4531 Bryan Avenue, Irvine, California, 92620

Site Type: Monument – Wars in Afghanistan and Iraq
Conflict: Afghanistan War, Iraq War
Dates: Dedicated in 2010 (periodically updated)
Website: http://northwoodmemorial.com (official website)

The Northwood Gratitude and Honor Memorial was the first monument erected in the United States to honor those who have died in Iraq and Afghanistan since 2001. It is updated periodically to accommodate additional names of those killed since the monument's completion in 2010.

The Northwood Memorial began in 2003 as a temporary exhibit displayed in Irvine over Memorial Day. Over the next few years the memorial was expanded as names were added, with special services on Veterans Day and Independence Day as well as candlelight vigils during the summer. In 2006 the city approved a permanent monument which was completed in 2010.

The Northwood Gratitude and Honor Museum consists of multiple pillars engraved with the names of the dead on a large open plaza. The style of the monument is reminiscent of the Vietnam Memorial in Washington DC. As of the time of this writing approximately seven thousand names were inscribed on the monument.

193. SAN DIEGO AIR AND SPACE MUSEUM

2001 Pan American Plaza, San Diego, California, 92101

Site Type: Museum
Dates: Opened in 1980
Web: www.sandiegoairandspace.org (official website)

The San Diego Air and Space Museum is the largest institution of its kind on the West Coast. Located in the old Ford Building on San Diego's exposition grounds, it is now part of the city's world famous Balboa Park museum district. Although not as large or as famous as the National Air and Space Museum in Washington, DC, it nevertheless has an excellent collection with aircraft from before World War I through the present day.

San Diego had an earlier air and space museum that opened in the 1960s. However, the museum and virtually its entire collection of planes and exhibits were destroyed in a fire in 1978. Some one-of-a-kind historic craft, including the Wee Bee, the smallest manned plane ever built, were lost. New planes were eventually acquired for the collection, but unfortunately some of the most famous planes were replaced only by replicas or reproductions. The San Diego Air and Space Museum became affiliated with the Smithsonian Institute in 2005.

The San Diego Air and Space Museum has largely recovered, with many new acquisitions. Large galleries showcase planes from World War I, World War II and the post-war era. The oldest non-replica plane on display is a Montgomery 1911 Evergreen Glider. The museum also has many exhibits about the inner workings of planes, including a collection of fuselage cutaways.

194. USS MIDWAY MUSEUM

910 North Harbor Drive, San Diego, California, 92101

Site Type: Aircraft Carrier
Conflicts: Vietnam War, Cold War, Iraq War
Dates: Commissioned in 1945
Web: www.midway.org (official website)

The USS Midway (CV-41) was the first aircraft carrier to go into service following the end of World War II, having received its commission barely a month after hostilities ceased in 1945. It was also the first of the Midway-class aircraft carriers that dominated the American navy in the early years of the Cold War. The Midway is perhaps most famous for its role in Operation Frequent Wind, when Americans and friendly civilians from South Vietnam were evacuated from Saigon in 1975. After Vietnam the Midway participated in the liberation of Kuwait during Operation Desert Storm in 1991.

The Midway went into service in 1945 and had what could be the most interesting history of any American post-war naval vessel. It saw service in the Atlantic in the 1940s and 1950s, making history when a rocket, a captured German V-2, was launched from a moving ship for the first time. It transferred to service in the Pacific in 1958.

The Midway did not see military action until the Vietnam War, but had several prominent moments during this conflict. Naval aviators from the Midway scored the first and last American air-to-air kills of the war. In 1975, during the fall of Saigon, helicopters from the Midway raced to evacuate as many American and friendly personnel from the capital as possible. Among the most enduring images of the Midway is that of helicopters being pushed off of its deck in a desperate attempt to speed up the evacuation. The USS Midway was decommissioned in 1992 and converted to use as a museum ship in 2004.

The USS Midway is enormous, being for a brief time the largest ship in the world, and is the only aircraft carrier museum in operation that was not in service during World War II. There is a small assortment of aircraft on display, including planes from the Cold War era. The USS Midway Museum is also located right next to the Maritime Museum of San Diego, which is home to a Soviet-era submarine and a number of other historic ships.

---

195. BOB HOPE MEMORIAL

Navy Pier, San Diego, California, 92132

Site Type: Monument – Bob Hope
Dates: Dedicated in 2009
Web: www.portofsandiego.org/wonderfront/embarcadero/tuna-harbor-park (official website)

The Bob Hope Memorial, officially known as *A National Salute to Bob Hope and the Military*, commemorates the most prolific entertainer of troops in American history. Beginning in World War II and continuing beyond the Vietnam conflict, Bob Hope put on countless live performances for tens of thousands of American servicemen around the world. By the time of his death he had received countless honors for his contribution to America's military esprit de corps.

Bob Hope was born in 1903, and like many entertainers of his generation got his start in vaudeville. He achieved Hollywood stardom at the height of World War II and became a popular regular participant of USO tours. Over the course of fifty years he did shows during World War II, the Korean War, the Vietnam War and the Gulf War. He continued performing for the troops well into his eighties.

Hope became a staple of the American military and political entertainment scenes. By the time of his death he had received hundreds of decorations including the Presidential Medal of Freedom and the Congressional Gold medal. In 1997 he was named an honorary veteran by Act of Congress. Bob Hope died in 2003.

The Bob Hope Memorial was dedicated in 2009. Located at San Diego's historic Navy Pier, this unique monument features a bronze statue of Bob Hope, microphone in hand, entertaining over a dozen bronze servicemen. The range of uniforms worn by the bronze statues reflects the changes over time of Hope's audiences.

---

196. USS MISSOURI MEMORIAL

63 Cowpens Street, Honolulu, Hawaii, 96818

Site Type: Battleship
Conflicts: World War II, Korean War, Gulf War
Dates: Commissioned in 1944
Designations: National Register of Historic Places
Web: www.ussmissouri.org (official website)

The USS Missouri (BB-63), or *Mighty Mo*, was possibly the most famous American battleship of World War II, and the last to see active service. One of the final battleships ever constructed in the United States, it was officially commissioned in 1944 just in time to participate in the final campaigns in the Pacific. Although it saw nearly continuous action during this period, the Missouri gained fame as the location where the Japanese formally surrendered, ending World War II. It later saw service in both the Korean War and the Gulf War. The Missouri was the last American battleship to ever fire its guns in combat.

The Missouri entered active service in the final year of World War II, supporting the landing invasions of Iwo Jima and Okinawa. Throughout the summer of 1945 it engaged in offshore shelling campaigns against targets on the Japanese home islands. After the atomic bomb attacks on Hiroshima and Nagasaki, the Missouri was chosen as the official site of the Japanese surrender. At the end of August she sailed into Tokyo Bay, and a few days later, On September 2, the Japanese signed the surrender with Admiral Chester Nimitz, General Douglas MacArthur and representatives of most of the major Allied powers present.

After World War II the Missouri probably had the most active post-war career of any battleship afloat. She participated in the landings at the Battle of Incheon in Korea in 1950, then continued to serve in a support position for much of the rest of the conflict. During the Gulf War the Missouri once again participated in an offshore bombardment campaign, the last time an American battleship ever did so. The USS Missouri received its final decommission in 1991. It was later moved to Pearl Harbor and refurbished as a museum ship in 1998. The USS Missouri was added to the National Register of Historic Places in 1971.

The USS Missouri Memorial is now located at Pearl Harbor's Ford Island, site of the heaviest fighting during the attack of December 7, 1941. In addition to the rooms on the ship which can be toured there are exhibits on the Korean War and on the Kamikaze attacks during the Battle of Okinawa. A plaque memorializes the location of the Japanese surrender in 1945.

---

197. USS ARIZONA MEMORIAL

1 Arizona Memorial Place, Honolulu, Hawaii, 96818

Site Type: Monument – Battle of Pearl Harbor
Conflict: World War II
Dates: Battle fought on December 7, 1941;
Monument dedicated in 1962
Designations: National Historic Landmark
Web: www.nps.gov/valr (official website)

The USS Arizona Memorial is the de-facto memorial site for the Battle of Pearl Harbor. Specifically commemorating the battleship Arizona, which was destroyed in the attack, the monument is located in the harbor directly over the sunken ship. It is one of several sites marking important places associated with the Japanese attack of 1941, but easily the most famous and important.

The surprise Japanese air attack on Pearl Harbor on December 7, 1941 is one of the most well-remembered dates in American history. It was, as President Franklin D. Roosevelt called it, the "day that will live in infamy", and heralded America's entry into World War II. It primarily involved an air assault by over four hundred carrier based Japanese aircraft on the unprepared American fleet at Pearl Harbor, as well as several American airfields around the island of Oahu. The attack, a complete success on the part of the Japanese, left the American naval and air forces in Hawaii in shambles.

The most famous casualty by far was the battleship USS Arizona. While all eight American battleships present during the attack were hit by Japanese bombs, the Arizona was only one of two, along with the Oklahoma, that were permanently sunk. One bomb hit the Arizona's powder magazine, resulting in an immense explosion that destroyed the ship and killed over a thousand men. The famous memorial was completed in 1962. It is part of the World War II Valor in the Pacific National Park site, which also has sites in Alaska and California. It was declared a National Historic Landmark in 1989.

The USS Arizona Memorial is itself a fascinating work of a modern architecture. Rising up out of the harbor, the visually distinctive white monument straddles the Arizona without touching it. There is a floor opening that allows visitors to look down upon the ship, a memorial to the men who died as well as to the handful of survivors of the disaster. The Arizona Memorial is only accessible by boat.

---

198. USS BOWFIN SUBMARINE MUSEUM

11 Arizona Memorial Drive, Honolulu, Hawaii, 96818

Site Type: Submarine
Conflict: World War II, Korean War
Dates: Commissioned in 1943
Designations: National Register of Historic Places, National Historic Landmark
Web: www.bowfin.org (official website)

The USS Bowfin is one of the last surviving World War II submarines of the Balao class. It was one of the largest submarines to see service during the war, and was active on and off for nearly three decades.

Launched exactly one year after the Japanese Attack on Hawaii, the Bowfin was nicknamed the *Pearl Harbor Avenger*. It is now permanently moored at Pearl Harbor where it serves as a museum ship.

The Balao class ships formed the backbone of America's submarine force during World War II. Over a hundred saw service during the conflict, and many survived the war. The Bowfin entered service in 1943 and ran nine patrols against Japan in the course of two years. It later served during the Korean War, then finally as a training vessel in the 1960s. The USS Bowfin was listed on the National Register of Historic Places in 1982 and named a National Historic Landmark in 1986.

The USS Bowfin was refurbished for use as a museum ship in 1979. Visitors can walk the decks of the submarine, but access to the interior is limited. The main museum building has exhibits on the history of the Bowfin, a collection of ship's objects and a Poseidon missile. The museum is also home to a memorial honoring those submarine crew members who lost their lives during World War II.

---

199. WAR IN THE PACIFIC NATIONAL HISTORICAL PARK

Guam Highway 1, Piti, Guam, 96931

Site Type: Battlefield – Battle of Guam
Conflict: World War II
Dates: Battle fought on July 21-August 10, 1944
Web: www.nps.gov/wapa (official website)

The Island of Guam, an American territory since 1898, was one of the first places annexed by the Japanese during World War II, and one of the last to be liberated. Guam has been home to several important American military facilities in the Pacific for over a century. It subsequently played a role in both the Korean and Vietnam conflicts.

Japan seized the island of Guam in December 1941, immediately after the surprise attack on Pearl Harbor. Due to its strategic importance the Japanese strongly fortified the island and held it for nearly three years. The Americans returned in the summer of 1944 and retook the island after three weeks of bitter fighting. Virtually all of the Japanese soldiers were killed during the battle.

The War in the Pacific National Historic Park was established on Guam in 1978. It commemorates all of those who fought in the Pacific War, including America's allies in the region. Trails through the park lead to a variety of landing beaches, defensive earthworks and gun emplacements.

---

200. AMERICAN MEMORIAL PARK

Micro Beach Road, Garapan, Northern Mariana Islands, 96950

Site Type: Monument – Marianas Campaign
Conflict: World War II
Dates: Dedicated in 1978
Web: www.nps.gov/amme (official website)

The Mariana and Palau Island Campaign was one of the most famous of the small-island campaigns undertaken by the United States in the Pacific during World War II. This massive strike at the Japanese possessions in the Central Pacific was a strategic disaster for the Axis, clearing the way for an American attack on the Philippines and setting the stage for air raids against the home islands.

The campaign began in June 1944 with the American capture of the island of Saipan. This was followed by one of the largest naval engagements of the war, the Marianas Turkey Shoot, where three of Japan's largest aircraft carriers were destroyed, along with approximately six hundred aircraft. This crippled the dwindling Japanese navy and allowed the Americans to seize the rest of the islands over the next few months.

American Memorial Park is a public park on the island of Saipan. In addition to the park's recreation facilities, there is a large monument honoring those who fought in the Marianas. The names of over five thousand American dead are inscribed on its granite blocks.

FOREIGN – NORTH AMERICA AND CARIBBEAN

201. MONUMENT TO THE VICTIMS OF THE USS MAINE

Malecon Boulevard & Linea Street, Havana, Cuba

Site Type: Monument – USS Maine
Conflict: Spanish-American War
Dates: Dedicated in 1925
Web: None Available

The Sinking of the USS Maine was a definitive historic moment for two countries: the United States and Spain. An event which helped to trigger the Spanish-American War, it led to the confirmation of America as a world power, and the final collapse of the four centuries old Spanish Empire in the New World. Ironically, the sinking of the Maine may have simply been an accident.

On February 15, 1898 the USS Maine, which had been sent to Cuba during that country's war of independence, mysteriously exploded in Havana Harbor. Although nothing was proven, war hawks in the United States, including several major journalists, blamed Spain, and used the rallying cry of "Remember the Maine" to drum up war against the Spanish. This contributed to America's declaration of war against Spain a few months later.

The Monument to the Victims of the USS Maine was erected in Havana in 1925 to commemorate the nearly three hundred sailors and officers who lost their lives in the disaster. It originally featured busts of William McKinley, Teddy Roosevelt and Leonard Wood, but these were removed in 1961 following the Communist revolution in Cuba. The memorial has since been repurposed as a propaganda tool, as its new name suggests two meanings.

202. SAN JUAN HILL BATTLEFIELD & SANTIAGO SURRENDER TREE

Avenida de Raul Pujols, Santiago de Cuba, Cuba

Site Type: Battlefield – Battle of San Juan Hill
Dates: Battle fought on July 1, 1898
Conflict: Spanish-American War
Web: www.abmc.gov/cemeteries-memorials/americas/santiago-surrender-tree (official website)

The Battle of San Juan Hill, or San Juan Heights, was the most storied battle of the Spanish-American War. Famous for the participation of future president Teddy Roosevelt, his Rough Riders and several regiments of the fabled Buffalo Soldiers, the Battle of San Juan Hill was the decisive engagement of the war in Cuba. The hard-won American victory left the strategically vital city of Santiago vulnerable and ultimately led to Spanish withdrawal from the island. The Santiago Surrender Tree marks the spot where Spanish forces capitulated to the Cubans and Americans at the end of the war.

In 1898 the sinking of the United States battleship Maine in Havana Harbor led to an American declaration of war against Spain. The war was waged around the world against the last few surviving colonies of the once mighty Spanish Empire. The prime target was Cuba, and a few months later an American army of about twenty thousand men was assembled near the key port city of Santiago.

The objective was the San Juan Heights which protected the approach to the city. Defended by less than a thousand Spanish soldiers, these were nevertheless well-entrenched and put up a ferocious defense. The American and Cuban forces assaulted the hill on July 1, and casualties on both sides were high. In the end, the overwhelming American forces carried the day. The Spanish suffered nearly one hundred percent casualties, though they killed or wounded over three thousand Americans in the defense. The battle led to the collapse of the Spanish government in Cuba. On July 17 the Spanish surrendered the island.

The San Juan Hill Battlefield actually consists of a small series of hills, the common use of the singular being attributed to a journalistic typo. Much of the battlefield has been preserved, including the American trenches, artillery pieces, and monuments to both the American and Cuban soldiers who participated in the battle. In 1958 the Santiago Surrender Tree was designated as an American monument. Technically under the jurisdiction of the American Battle Monuments Commission, the tree is currently maintained by the Cuban government. The original tree was replaced in 1998 on the one hundredth anniversary of the battle.

---

203. CHAPULTEPEC PARK

Avenue Paseo de la Reforma, Miguel Hidalgo, Mexico City, Mexico

Site Type: Battlefield – Battle of Chapultepec
Conflict: Mexican-American War
Dates: Battle fought on September 12-13, 1847
Web: https://mnh.inah.gob.mx
(official tourism website of Mexico City)

The Battle of Chapultepec was the last major engagement of the American campaign to capture Mexico City during the Mexican-American War. It was followed by the storming of the city, which crippled the Mexican war effort and effectively ended the imperial government of Santa Ana. Although an American victory, the battle is remembered for heroism on both sides. One of Mexico's greatest war monuments is here, commemorating the young cadets who fought valiantly defending the city to the death.

The Mexican-American war was the largest conflict between two sovereign, non-European nations in the Americas. Although many European military observers expected a decisive Mexican victory, the war consisted largely of a series of defeats for Mexico. By the second half of 1847 the Americans had already penetrated deep into the country. Following engagements at Contreras, Churubusco and Molino del Rey, the Mexican army was driven back into Mexico City.

Its last surviving forward outpost was Chapultepec Castle which was garrisoned by a force of Mexican regulars as well as detachment of about two hundred cadets, some little more than children. For two days the Americans bombarded the fortress. On the second day the Americans assaulted Chapultepec with scaling ladders and took the outside wall. The commander of the garrison then ordered a withdrawal back to Mexico City. However, six cadets, all teenagers, famously refused the order, and proceeded to fight to the death. The battle inspired two famous literary works: one which later became the *American Marine Corps Hymn*; and *Los Ninos Heroes*, a popular Mexican poem honoring the six cadets.

Chapultepec Park is home to one of Mexico's best and most interesting battlefield sites. The castle was rebuilt after the war, and in 1864 it was briefly made the residence and court of the Mexican imperial family. Outside the castle, in the general area where the American assault took place, is the Monument to the Boy Heroes which commemorates the young cadets who died in defense of the capital. Both the castle and battlefield are preserved in the sprawling Chapultepec Park.

---

204. MEXICO CITY NATIONAL CEMETERY

31 Virginia Fabregas, Colonia San Rafael, Mexico City, Mexico

Site Type: Cemetery
Conflict: Mexican-American War
Dates: Established in 1851
Web: www.abmc.gov/cemeteries-memorials/americas/mexico-city-national-cemetery (official website)

The Mexico City National Cemetery is not, as its name implies, the state cemetery of the nation of Mexico. It is actually a cemetery established by the United States Congress a few years after America's victory in the Mexican-American War. Still maintained as an American memorial site, the majority of those buried here are unidentified.

The Mexican-American War was one of the largest military campaigns waged by the United States prior to the Civil War. Despite their victory, approximately fifteen thousand Americans were killed during the conflict, many while campaigning in Central Mexico. After the war, land was acquired to bury hundreds of unnamed dead, with a cemetery formally established in 1851. It is now run by the American Battle Monuments Commission.

The Mexico City National Cemetery is small, about one acre, but over eight hundred American soldiers are interred here. Most of these are unknown and buried in a common grave. However there are a few marked gravesites, including that of General James Slaughter, who served in both the Mexican-American War and the American Civil War.

---

205. FORT MALDEN NATIONAL HISTORIC SITE

100 Laird Avenue South, Amherstburg, Ontario, N9V 1W4, Canada

Site Type: Wooden Star Fort
Conflict: War of 1812
Dates: Originally completed in 1797
Designations: National Historic Site of Canada
Web: www.pc.gc.ca/en/lhn-nhs/on/malden (official website)

Fort Malden, officially known as Fort Amherstburg, is an historic Canadian military fort that has the distinction of having once been under American occupation. Founded in the late 19th century, Fort Malden was established to protect Ontario against the threat of an American invasion from Michigan. It was fought over by the Americans and British during the War of 1812, and served for a time as a base for the warriors of the Tecumseh Confederacy.

After the American Revolution, the British, concerned about future hostilities with the United States, built Fort Malden to guard the southern entrance to the Detroit River. The area around the fort became a focal point for native tribes fleeing American authority. Some of the earliest fighting of the War of 1812 took place near Fort Malden. The fort was destroyed and abandoned in 1813, then subsequently occupied by the American Army for the rest of the war.

Fort Malden once again played a small role in American history in the years leading up to the Civil War. It was one of the major terminus locations in Canada for the Underground Railroad, and thousands of escaped African slaves finally reached sanctuary and freedom when they arrived here. The fort was abandoned in 1870, after which much of it was sold off. Fort Malden was named a Canadian National Historic Site in 1921.

Much of Fort Malden is now gone, but some of the earthworks and historic buildings still remain. The main points of interest are the partially recreated stockade and the surviving barracks, the latter of which dates from the early 19th century. The visitor's center houses a small museum with exhibits on the history of the fort.

---

206. THAMES BATTLEFIELD

Moraviantown, Chatham-Kent, Ontario, N0P 1C0, Canada

Site Type: Battlefield – Battle of the Thames
Conflict: War of 1812
Dates: Battle fought on October 5, 1813
Web: www.chatham-kent.ca/tourism
(official tourism website of Chatham-Kent)

The Battle of the Thames was one of the decisive engagements of the War of 1812, and one of the few major American victories of that conflict. Following the brutal defeat at the Battle of River Raisin nine months earlier, the Battle of the Thames allowed the United States to retake Detroit and secure the northwestern frontier. The battle was famous for two men who fought there: William Henry Harrison, who led the Americans; and Tecumseh, who led the Shawnee Confederacy in alliance with the British.

Early in 1813 the British had gained the upper hand in Michigan after their victory at Raisin River in January. However, with the defeat of the British naval force at the Battle of Lake Erie in September, the British army in Michigan could no longer be supplied and was forced to retreat into Ontario. The American army, which had grown significantly in the interim, pursued the British and their native allies.

The Americans caught up with the British at the banks of the Thames, and the British turned to make a stand. The fighting did not last long. Most of the British surrendered quickly, though Tecumseh and his men fought on ferociously. Tecumseh was killed during the fighting, which came to an end shortly thereafter. Although casualties were relatively light, almost all of the British force was captured. The battle made a hero out of Harrison, who went on to become the president, and contributed to the legend of Tecumseh.

The Thames Battlefield is not a formally preserved site, though markers note where the battle occurred. Much of the original village was destroyed by the occupying American forces, though there is a replica of an original period cabin on the battle site.

FOREIGN – NORTHERN EUROPE

207. AMERICAN AIR MUSEUM IN BRITAIN

106 A505, Cambridge, CB22 4QR, United Kingdom

Site Type: Museum
Dates: Opened in 1977
Web: www.iwm.org.uk/visits/iwm-duxford (official website)

The American Air Museum in Britain is part of the Imperial War Museum Duxford. This branch of the Imperial War Museum is home to exhibits on military aviation with a focus on the World Wars. Because the American and British air commands were so closely integrated for joint operations during World War II, the museum became home to a collection of American aircraft that participated in the war effort.

The Imperial War Museum was originally founded in 1917 at the height of World War I. The aerodrome at Duxford was founded the same year to train pilots for that conflict. Duxford was loaned to the American Army Air Corps during World War II as a fighter base. After the war Duxford became a home to refurbished war planes and was a popular location for air shows. It was formally incorporated into the Imperial War Museum in 1977.

The IWM Duxford consists of six large buildings, including original hangars, full of largely World War II aircraft. The American Air Museum was added in the 1980s and is located in its own dedicated building. Jimmy Doolittle was one of the project's early supporters. The room is packed with American aircraft from both World Wars, notably three super bombers: a B-17 Flying Fortress, a B-29 Superfortress and a B-52 Stratofortress.

208. LUSITANIA MEMORIAL GARDEN

Downmacpatrick, Kinsale, County Cork, Ireland

Site Type: Monument – Sinking of the Lusitania
Conflict: World War I
Dates: Dedicated in 2017
Web: www.kinsale.ie/2019/03/04/lusitania-memorial-garden (official tourism website of Kinsale)

The Lusitania Memorial Garden commemorates the RMS Lusitania, a turn of the century ocean liner that became famous for its sinking by German U-boats in 1915. Although a civilian ship, the Lusitania was collateral damage of the war and went down with nearly twelve hundred passengers killed, including over a hundred Americans. While this act did not drag the United States immediately into World War I, it did turn the tide of American opinion against Germany and contributed to their declaration of war two years later.

The Lusitania was one of the great ocean-going passenger ships of the early 20th century. Launched in 1906, it made regular trans-Atlantic runs from Liverpool to New York City. With the outbreak of World War I, neutral and civilian shipping around the British Isles became dangerous due to the threat of German U-boats. The German embassy in Washington issued a warning to Americans in April 1915 about the dangers posed by the state of war between Great Britain and Germany.

On May 17, 1915 a German U-boat crossed paths with the Lusitania off the coast of Ireland, and without warning sent a single torpedo into the ocean liner. The Lusitania went down in under twenty minutes. Only a handful of lifeboats were launched, and only 764 of the 1,962 passengers on board survived. 128 Americans were killed. The deaths of the Americans sent a shockwave through the United States, stoking anti-German sentiment that would later lead the two countries to war.

The Lusitania Memorial Garden stands on the Old Head of Kinsale Peninsula, not too far from where the Lusitania was sunk.

According to some sources, the sinking was witnessed by onlookers near here. Several monuments are located in the gardens, including a large photo memorial which, when viewed from a certain angle, looks as if the Lusitania is just offshore.

---

209. UNITED STATES NAVAL MONUMENT AT BREST

Avenue Salaun Penquer, Brest, 29200, France

Site Type: Monument – American Navy
Conflict: World War I
Dates: Dedicated in 1958
Web: www.abmc.gov/cemeteries-memorials/europe/naval-monument-brest (official website)

The city of Brest In Brittany in northwestern France was the head-quarters of the American Navy in Europe during World War I and the primary destination for American troop and supply ships supporting the war effort. More than a third of all American servicemen who fought in the war passed through here, along with countless cargoes of armaments, munitions, food and other supplies heading for the Western Front.

In addition to its commercial importance, Brest was critical as a base for American and French naval ships hunting down German submarines. After the war France commissioned the monument to honor the contribution of the navies of both countries.

The current United States Naval Monument at Brest, which stands by the ramparts of the old medieval city walls, dates from the 1950s. It is a close replica of the original monument, which was destroyed by the Nazis in 1941 during their occupation of France in World War II. The tower, which is reminiscent of an old Moorish minaret, actually stands on top of a former German World War II bunker.

210. NORMANDY AMERICAN CEMETERY AND MEMORIAL

Normandy American Cemetery, Colleville-sur-Mer, 14710, France

Site Type: Battlefield – Battle of Normandy/D-Day Landings
Conflict: World War II
Dates: Battle fought on June 6, 1944
Web: www.abmc.gov/cemeteries-memorials/europe/normandy-american-cemetery (official website)

The Battle of Normandy was the largest seaborne invasion in world history. Thousands of ships and planes supported the landing of hundreds of thousands of soldiers in what was arguably World War II's most storied battle. The success of the amphibious assault was followed by the liberation of Northern France and ultimately the collapse of the Third Reich in the west. Of the many places along the Normandy coast associated with the battle, perhaps the most famous and visited is Omaha Beach, site of the bloodiest fighting on D-Day.

Operation Overlord, the code name for the campaign, had been meticulously planned over the course of two years, and by May 1944 the Allies had assembled the largest amphibious invasion force in history in Southern England. Over a million soldiers from more than a dozen countries were gathered under the command of Dwight Eisenhower, along with over a hundred thousand vehicles, thousands of warships and landing craft, and countless airplanes. Opposing them along the northern coast of France were hundreds of thousands of heavily armed, well entrenched German soldiers led by veteran commanders Von Rundstedt and Rommel.

The primary target, at least for the Americans, was Omaha Beach. The battle began with airborne units securing vital rear areas and opening the door for the subsequent amphibious assault. The landings followed just after dawn, with tens of thousands of men arriving in wave after wave to attack the vaunted Atlantic Wall. It was at Omaha that the Germans put up the fiercest resistance, and where American casualties were the highest. Despite this Omaha was taken on the first day, along with the rest of the beaches, and most of the Allied gains were linked up and consolidated within a week.

The Normandy American Cemetery and Memorial is located just inland from Omaha Beach, and is the most popular Normandy destination for American visitors. In addition to the memorial building, the graves of nearly ten thousand American soldiers can be found here, many of whom were killed on June 6. Beyond the cemetery visitors can look down on the beach and some of the surviving German fortifications.

---

211. UTAH BEACH AMERICAN MEMORIAL & MUSEUM

Plage de la Madeleine, Ste-Marie-du-Mont, 50480, France

Site Type: Battlefield – Battle of Normandy/D-Day
Conflict: World War II
Dates: Battle fought on June 6, 1944
Web: www.abmc.gov/cemeteries-memorials/europe/utah-beach-american-memorial (official website)

Utah Beach was one of two combat zones, along with Omaha Beach, that were assaulted and taken by American troops during the D-Day landings at Normandy. The westernmost end of the assault zone, the attack at Utah was designed to secure the right flank of the invasion and set the stage for the capture of the port of Cherbourg, one of the immediate objectives of the campaign. Although perhaps less storied than the assault on neighboring Omaha, it was no less critical to achieving the victory at Normandy.

The battle at Utah Beach was the location of some of biggest successes on June 6. The landings were personally led by Teddy Roosevelt, Jr., son of former president Theodore Roosevelt. Significantly more territory was captured here on D-Day than at Omaha, and in less than a week a sizeable piece of the Cherbourg Peninsula was taken from

Utah's expanding beachhead. The 82nd Airborne landed inland from Utah where they had their famous engagement with the Germans at the town of Sainte-Mere-Eglise.

There are two sites of interest here: the Utah Beach American Memorial, which overlooks the site of the landings, and the Utah Beach Museum. The memorial consists primarily of a monument dedicated to the VII Corps, who shouldered much of the fighting here for the better part of a month. The relatively new Utah Beach Museum is home to extensive exhibits on the history and artifacts of the battle.

---

212. MEMORIAL MUSEUM OF THE BATTLE OF NORMANDY

Boulevard Fabian Ware, Bayeux, 14400, France

Site Type: Museum
Conflict: World War II
Dates: Opened in 1981
Web: www.bayeuxmuseum.com/en/memorial-museum-battle-of-normandy (official website)

The Museum of the Battle of Normandy is France's official museum documenting the events of the Normandy Invasion of 1944. Opened in 1981, this museum has exhibits not only the famous landings of June 6, but the ensuing three month struggle to liberate Paris. It is among the finest museums of military interest in Europe.

On June 7, 1944, the day after the Normandy invasion began, Bayeux was the first major French city to be liberated. Just inland from the British beach zone known as Gold, Bayeux served as an important British command center in the days immediately following the D-Day landings. It remained so until the Allies took Caen and more extensively secured the Normandy beachhead.

The Memorial Museum of the Battle of Normandy is a large building absolutely packed with objects from the Normandy landings and the subsequent fighting in Northern France. Many artifacts from the battle can be found here, from guns to equipment to uniforms, almost all of which has been restored and/or preserved. There is also a film that briefly narrates the history of the battle.

---

213. TOURS AMERICAN MONUMENT

117 Avenue Andre Malraux, Tours, 37000, France

Site Type: Monument – Services of Supply
Conflict: World War I
Date: Dedicated in 1937
Web: www.abmc.gov/cemeteries-memorials/europe/tours-american-monument (official website)

The United States did not enter World War I until April 1917, and did not reach full military strength in France until close to the end of the war in November 1918. During that period of 19 months, United States forces had to develop a massive infrastructure in France in order to support their campaign.

The Services of Supply numbered well over half a million men, more than a quarter of the total troops committed to the war in France, plus over twenty thousand civilian support staff. In barely a year and a half they accomplished a staggering number of projects, including everything from the expansion of the French rail network to the establishment of hospitals with a capacity to treat hundreds of thousands of wounded soldiers.

The Tours American Monument commemorates the efforts and accomplishments of the Services of Supply. The monument, in the form of a large public fountain, is crowned with a gold statue of an American bald eagle being held up by a Native American. It is maintained by the American Battlefield Monuments Commission.

214. CHAUMONT AMERICAN EXPEDITIONARY FORCE HEADQUARTERS MARKER

1 Avenue du 109eme RI, Chaumont, 52903, France

Site Type: Monument – World War I Headquarters
Conflict: World War I
Dates: Dedicated in 1923 (?)
Web: www.abmc.gov/cemeteries-memorials/europe/chaumont-marker-aef-headquarters (official website)

For nearly two years, from the autumn of 1917 through the summer of 1919, an old mansion in Chaumont served as the headquarters of the American Expeditionary Force in France. It was from here that John J. Pershing directed the activities of millions of American servicemen in the First World War. A marker commemorates the site more than a century later.

In April 1917 the United States declared war against Germany, becoming the last of the great powers to enter World War I. Pershing was named the supreme commander of the American Expeditionary Force. He arrived in France in May, followed by the first American troops in June. In September Pershing established his command at the Caserne Damremont, a building formerly used as a barracks in the city of Chaumont. This remained the AEF headquarters from then until the summer following the end of the war. The building was turned back over to the city in July 1919.

More than a century later the Caserne Damermont, where the American Expeditionary Force was based, is still standing in Chaumont. Although lesser known and less visited than other American military sites in France, a large plaque on the front of the building commemorates the former AEF headquarters which once occupied the building. It is maintained by the American Battle Monuments Commission.

215. LAFAYETTE ESCADRILLE MEMORIAL CEMETERY

5 Boulevard Raymond Poincare, Marnes-la-Coquette, 92430, France

Site Type: Monument – Lafayette Escadrille
Conflict: World War I
Date: Dedicated in 1928
Web: www.abmc.gov/cemeteries-memorials/europe/lafayette-escadrille-memorial-cemetery (official website)

The Lafayette Escadrille was a squadron of pilots from the United States who volunteered to fight for France prior to America's entry into World War I in 1917. Part of a larger volunteer movement that also saw Americans go to France to help in a number of capacities, approximately two hundred American pilots fought for France from 1915 to 1917. France later established a cemetery and memorial arch in honor of their heroics.

American pilots were flying for France as early as 1915. These volunteers served as part of existing French squadrons. In 1916 the French government authorized the creation of an American squadron which came to be known as the Lafayette Escadrille in honor of the Marquis de Lafayette, a French general who aided the patriots during the American Revolution. While not all of the American flyers were part of this unit, they were all considered part of the greater Lafayette Flying Corps.

The Lafayette pilots fought in some of the largest actions of the war, especially in 1916. Among these were the battles of the Somme and Verdun. The American squadron was disbanded after America entered the war in 1917, and its pilots transferred to service in the Air Corps. Some of the Lafayette pilots served after the war in actions in the French colonies. The Lafayette Escadrille Memorial Cemetery was established in 1928.

The Lafayette Escadrille Memorial Cemetery includes graves for all of those American flyers who died in the service of France, though not all are buried here. The centerpiece of the cemetery is a magnificent triumphal arch, one of the largest in France. It was recently completely renovated in honor of the 100th anniversary of the formation of the squadron.

---

216. BELLEAU WOOD AMERICAN MONUMENT

Route de Torcy, Belleau, 02400, France

Site Type: Battlefield – Battle of Belleau Wood
Conflict: World War I
Dates: Battle fought on June 1-26, 1918
Web: www.abmc.gov/cemeteries-memorials/europe/belleau-wood (official website)

The Battle of Belleau Wood was possibly the largest engagement fought primarily by United States Marines, as opposed to the Army, in Europe during World War I. In the engagement, part of the German offensive of 1918, the Marines distinguished themselves in some of the most ferocious fighting by American units during the war.

By June of 1918 the juggernaut known as the German Spring Offensive threatened to finally break through the Allied positions and seize Paris. One of the areas of greatest threat was at the Belleau Wood where, on June 1, German soldiers overcame a part of the French line. In response, nearby units of the American second division were rushed to stop the German advance.

Marine units attached to the second division distinguished themselves in the counterattack. The fighting over the next few weeks was intense, and the marines took heavy casualties. It took them nearly a month to beat back the Germans, often in hand-to-hand combat conditions. The forest was finally retaken on June 26th. A few days later, it was renamed the Wood of the Marine Brigade in honor of the heroic defense.

The Belleau Wood American Monument is located in the center of a small forest a short distance from the Aisne-Marne American Cemetery. At the heart of the woods is a large bronze monument commemorating the marines who fought here. Also nearby are remnants of trenches from the battle as well as artillery pieces.

---

217. CHATEAU-THIERRY AMERICAN MONUMENT

Route du Monument, Chateau-Thierry, 02400, France

Site Type: Battlefield – Second Battle of the Marne
Conflict: World War I
Dates: Battle fought on July 15-August 6, 1918
Web: www.abmc.gov/cemeteries-memorials/europe/chateau-thierry-monument (official website)

The Second Battle of the Marne was part of the greater German Offensive of 1918. It was also one of the most important defensive campaigns that involved the use of large numbers of American troops. Fought through much of the summer, the Americans distinguished themselves in the fight to prevent Germany's last desperate attempt to capture Paris.

Although the United States entered World War I in 1917, it wasn't until the summer of 1918 that they had sufficient force to fight independently in France. Many divisions were just being deployed at the front when the German summer offensive began on July 15. The arrival of tens of thousands of American soldiers helped to stem the tide. By the time the battle was over the Germans had been pushed back and America had taken over ten thousand casualties, the highest toll in any single battle since the Civil War.

The Chateau-Thierry American monument was erected in 1937. A huge stone monument whose façade recalls the Lincoln Memorial in Washington commemorates both American and French soldiers who fought side by side in this area during World War I, primarily at the Second Battle of the Marne. A recently opened visitor center displays artifacts from the battle.

218. BELLICOURT AMERICAN MONUMENT

D1044, Bellicourt, 02420, France

Site Type: Battlefield – Battle of St. Quentin Canal
Conflict: World War I
Dates: Battle fought on September 29-October 10, 1918
Web: www.abmc.gov/cemeteries-memorials/europe/bellicourt-american-monument (official website)

The Battle of St. Quentin Canal was one of the pivotal engagements of the Hundred Days Offensive that culminated in Germany's surrender in World War I. The allied objective of the battle was to get across the canal and secure the Bellicourt Tunnel in order to move tanks into enemy territory. A large number of American troops fought here under British leadership. The first major allied penetration of the Hindenberg line occurred here.

The attack began on September 29 with one of the largest artillery barrages of the war. Early American assaults on the German positions did not go well, but British and Australian troops managed to stage an unexpectedly successful attack that led to the surrender of many of the enemy. By the end of the fighting the Allies had successfully secured a foothold on the far side of the canal.

The Bellicourt Monument was raised to honor the Americans who fought here and who, despite fielding only two of thirty-two divisions, sustained the heaviest casualties. The monument, which looms over the hard-won canal tunnel, is not too far from the Somme American Cemetery and Memorial.

219. MEUSE-ARGONNE AMERICAN MEMORIAL & MONTFAUCON AMERICAN MONUMENT

Rue d'Amerique, Montfaucon-d'Argonne, 55270, France

Site Type: Battlefield – Battle of the Meuse-Argonne
Conflict: World War I
Dates: Battle fought on September 26-November 11, 1918
Web: www.abmc.gov/cemeteries-memorials/europe/meuse-argonne-american-cemetery (official website)

The Meuse-Argonne Offensive was the last major military campaign of World War I, as well as the battle in which the American Expeditionary Force made its largest contribution to the Allied effort. Fought from the end of September until the signing of the Armistice on November 11, this massive push included more than a million fresh American troops and resulted in the collapse of a significant portion of the German army.

One of the most decisive actions of the Meuse-Argonne Offensive was the Battle of Montfaucon, which took place from October 14-17. After three weeks of hard fighting, it was at Montfaucon that the American Expeditionary Force had its big breakthrough. American troops penetrated the Hindenberg Line here, opening the way for the capture of the Argonne Forest and threatening the vital German supply center at Sedan.

The Meuse-Argonne American Memorial and Montfaucon American Monument both honor those who fought and fell in the Meuse-Argonne and related campaigns, and include a sprawling cemetery for the thousands of American soldiers who died here. The monument at Montfaucon was erected in 1937. The immense granite column is crowned with an observation deck that affords a good view of the surrounding area and part of the battlefield.

220. LUXEMBOURG AMERICAN CEMETERY AND MEMORIAL

50 Val du Scheid, 2517 Luxembourg City, Luxembourg

Site Type: Cemetery
Conflict: World War II
Dates: Established in 1951
Web: www.abmc.gov/cemeteries-memorials/europe/luxembourg-american-cemetery (official website)

The Luxembourg American Cemetery and Memorial is mainly a monument to those American soldiers who died fighting at the Battle of the Bulge during World War II. Over five thousand soldiers who died during the famous German counteroffensive are buried here.

On December 16, 1944, Germany initiated the Ardennes Counter-offensive, which rolled across recently liberated Luxembourg. It took nearly a month of fierce fighting for the Allies to drive the Germans back again. During that month the Americans, who bore the brunt of the fighting, saw some of the heaviest casualties of the war. The American cemetery in Luxembourg was in use in December of 1944 before the fighting had even come to a halt.

The Luxembourg American Cemetery was formally established in 1951. Most of the gravestones are marked, but there are graves of over a hundred unknown soldiers here as well. Undoubtedly the most famous burial here is that of General George Patton, who commanded the American Third Army and was pivotal in organizing the Allied counterattack.

221. BASTOGNE WAR MUSEUM & MARDASSON MEMORIAL

Colline du Mardasson 5 & Route de Bizory 1, 6600 Bastogne, Belgium

Site Type: Battlefield – Siege of Bastogne
Conflict: World War II
Dates: Siege fought from December 16, 1944-January 25, 1945
Web: www.bastognewarmuseum.be (official website)

The Battle of the Bulge was the last major German offensive of World War II. Personally planned by Adolph Hitler, it was a desperate gamble to dislodge the American and British armies approaching the Rhineland and regain the initiative along the western front. Despite fielding their last viable offensive army, the Germans were soundly defeated after making only small progress towards their strategic objective, the port of Antwerp. This disaster sealed the fate of the Third Reich, which surrendered unconditionally a few months later.

In the second half of 1944 Allied armies made their way inexorably towards Germany, liberating most of France and parts of the Low Countries by the end of the year. As the Americans approached the Rhine, Hitler personally ordered a counteroffensive set for the winter. Despite the strenuous objections of his general staff, the offensive was carried out, and on December 16 they struck. The Allies were not immediately prepared for the sudden German advance into Luxembourg and Belgium, and over the course of the first few days the Wehrmacht advanced about sixty miles before grinding to a halt.

Although the major reason for the stalled offensive was the overwhelming forces arrayed against them, the immediate reason for the end of the German advance was the failure to take the city of Bastogne. The Siege of Bastogne became the epicenter of the battle. Held only by the tenacity of the American 101st Airborne division, the Germans spent five days trying to take the city. However, an American counteroffensive led by George Patton relieved Bastogne and blunted the German campaign. Within a few weeks it was all over. Nearly half of the German soldiers were casualties, and the survivors retreated to the Siegfried line.

Two locations in Bastogne commemorate the Battle of the Bulge: the Bastogne War Museum and the Mardasson Memorial. The museum is home to exhibits and artifacts on World War II with a focus on the Siege of Bastogne. The latter is a great stone and marble monument honoring the American soldiers who died defending the town. Additionally, remnants of the siege and battlefield can be found in locations scattered around the city.

---

222. ARDENNES AMERICAN CEMETERY AND MEMORIAL

164 Route du Condroz, 4121 Neupre, Belgium

Site Type: Cemetery
Conflict: World War II
Dates: Opened in 1945
Web: www.abmc.gov/cemeteries-memorials/europe/ardennes-american-cemetery (official website)

The Ardennes American Cemetery and Memorial was established while World War II was still raging in Europe. While many of those buried here died during the Battle of the Bulge, there are also those who fought in various actions throughout the Low Countries and during the American Army's drive towards the Rhine River.

The Ardennes Forest is probably more famous for the fighting that took place in the area during World War I. However, some of the fiercest fighting of the Battle of the Bulge in World War II occurred here as well. After the Battle of the Bulge, the area became a staging point for American soldiers driving on the German city of Aachen.

The Ardennes American Cemetery and Memorial was formally established in the 1960s. However, Americans had been buried here since February 1945, after the collapse of the German offensive. Eventually over five thousand servicemen were interred here. In addition to the soldiers that fought at the Ardennes, the bodies of hundreds of downed airmen, many recovered after the end of the war, were buried here as well.

---

223. NETHERLANDS AMERICAN CEMETERY AND MEMORIAL

Amerikaanse Begraafplaats 1, 6269 NA Margraten, Netherlands

Site Type: Cemetery
Conflict: World War II
Dates: Opened in 1944
Web: www.abmc.gov/cemeteries-memorials/europe/netherlands-american-cemetery (official website)

The Netherlands American Cemetery and Memorial is the most prominent site of American military interest in the Netherlands. It is located at the extreme southern end of the country, the only part of the Netherlands where Americans fought during World War II.

The site where the American Cemetery now stands saw significant military activity throughout the war. A road through the area was used by the Nazis both in their advance against France in 1940 and their retreat back to Germany in 1945. During the Allied advance the dead of several different armies were buried here, including Germans.

The Netherlands American Cemetery and Memorial was first used in 1944. After the war large numbers of the corpses were exhumed and removed to their home countries, including many fallen American soldiers. However, more than eight thousand Americans remained buried here, and the cemetery as it currently exists was dedicated in 1960.

224. HURTGEN FOREST BATTLEFIELD

Brandenburger Tor, 52393 Hurtgenwald, Germany

Site Type: Battlefield – Battle of Hurtgen Forest
Conflict: World War II
Dates: Battle fought on September 19-December 16, 1944
Web: https://liberationroute.com/germany/huertgenwald (official website)

The Battle of Hurtgen Forest was one of the largest campaigns of World War II fought by American soldiers on German soil. There were several objectives of the battle, including support for the attack on the city of Aachen, securing part of the Siegfried line and taking the Rur Dam. Undertaken while the Germans were still capable of mounting a defense, the battle was difficult, casualties on both sides were high, and ultimately it was called off.

Barely three months after the landings at Normandy the Allies had liberated Paris and pushed across eastern France and into Belgium. By September forward elements were already in Germany and moving on the city of Aachen. Seeking to secure as much German territory west of the Rhine River as quickly as possible, the Americans launched an offensive against the Siegfried line in the Hurtgen Forest area.

The forest was much more heavily defended than expected. The German army was thoroughly dug in, and an extensive defensive network had been prepared. Fighting was fierce, and though the Americans slowly gained ground, the Germans held out long enough and inflicted enough casualties that when the Battle of the Bulge began the Americans were forced to withdraw. Considered a defensive victory for Germany, it was only temporary, as the American army conquered most of the area just a few months later.

The Hurtgen Forest Battlefield is dotted with memorials and other points of interest. Most of these are monuments and markers that commemorate the Americans who fought here, and who suffered some of the highest casualties for United States forces in the invasion of Germany. Also throughout the forest are clusters of gravesites, as well as the remains of German defensive structures and bunkers.

225. PEACE MUSEUM BRIDGE AT REMAGEN

An der Alten Rheinbrucke 11/Rheinpromenade,
53424 Remagen, Germany

Site Type: Battlefield – Battle of Remagen
Conflict: World War II
Dates: Battle fought on March 7-25, 1945
Web: www.bruecke-remagen.de (official website)

The Battle of Remagen was one of the most unanticipated and intense military engagements on the western front during the last months of the war in Europe. The target for both sides was the Ludendorff Bridge, better known as the Remagen Bridge, the only bridge across the Rhine River to be taken intact by the Allies. Its capture sped up the Allied timetable for the invasion of Germany and probably shortened the war by several weeks.

By March of 1945 the German defenses against the Allies in the west were collapsing quickly. The one great remaining natural obstacle was the Rhine River. Over the course of several months the Germans systematically destroyed all of the bridges crossing the Rhine in order to slow the Allied advance. However, the demolition of the Ludendorff Bridge did not go as planned, and on March 7 the American army grabbed it largely still intact.

The bridge, built during World War I by Germany as a means to support the war in France, was especially strong. For the next ten days the Germans tried desperately to destroy it, during which time the Americans got six divisions across. The bridge finally collapsed on March 17, by which time it was too late for the Germans. Never rebuilt, the capture of the Remagen Bridge has become one of the war's great legends.

The Peace Museum Bridge at Remagen is located in and around the surviving towers of the bridge. The rest of the bridge, which was deemed unnecessary to rebuild after the war, was ultimately demolished. The museum houses exhibits on the history of the bridge, the battle, and the German prisoners of war that were kept in camps in Remagen during the last few months of the war.

---

226. ALLIED MUSEUM

Clayallee 135, 14195 Berlin, Germany

Site Type: Museum
Conflict: Cold War
Dates: Opened in 1998
Web: www.alliiertenmuseum.de (official website)

The Allied Museum in Berlin is a gallery documenting the political and military history of Berlin during and immediately after the Cold War. Named for the Allies who occupied the western sectors of the city in the postwar era, the museum commemorates the struggle of the United States, Great Britain and France to keep West Berlin free of Soviet domination from 1945 until 1989.

After the final surrender of Germany at the end of World War II, the major allies in the European theater divided up Germany, and Berlin, into zones of occupation. The Soviet zone later became East Germany with its capital at East Berlin. The American, British and French zones combined to form West Germany, with an isolated outpost at West Berlin. For over four decades Berlin was the location of considerable political and diplomatic intrigue between the two sides.

In order to preserve West Berlin's independence, the Western Allies, in particular the United States, kept a strong military presence in the city. In addition to being a deterrent to the Soviets, the American force in West Berlin participated in a number of historic events of the Cold War, most notably the Berlin Airlift in 1948. The American military presence in Germany was drawn down significantly with the end of the Cold War.

The Allied Museum was opened in 1998 in an area of Berlin that was formally home to the United States garrison. There are exhibits on the initial occupation of the city at the end of the war, the Berlin Airlift, Checkpoint Charlie, and life in the American sector. Among the museum highlights is a transport plane used during the Berlin Airlift and the remnants of an old spy tunnel.

---

227. CHECKPOINT CHARLIE MUSEUM

Friedrichstrasse 43-45, 10969 Berlin, Germany

Site Type: Museum
Conflict: Cold War
Dates: Opened in 1963
Web: www.mauermuseum.de (official website)

Checkpoint Charlie was arguably the most famous guardhouse in American military history, and an iconic site of the Cold War. Located at the most popular crossing point between the American and Soviet sectors of post-war Berlin, it became the stuff of legend once the Berlin Wall was erected in the 1960s. The checkpoint was manned by American soldiers until the wall came down in the 1990s.

A number of incidents occurred at Checkpoint Charlie over the years, including a tense showdown between American and Russian tanks in 1961. There were also several famous dashes for freedom through the checkpoint in the early days of the Berlin Wall. The wall was mostly leveled in 1989, and the Checkpoint Charlie guardhouse was removed a year later.

The Checkpoint Charlie Museum was created after the reunification of Germany and is home to exhibits and memorabilia related to the checkpoint and Berlin during the Cold War. A replica of the guardhouse is on display nearby, along with recreations of the famous *You are leaving the American Sector* signs. The original guardhouse is on display at the Allied Museum, also in Berlin.

---

228. MUSEUM OF THE PRISONER OF WAR CAMPS

Lotnikow Alianckich 45, 68-100 Zagan, Poland

Site Type: POW Camp
Conflict: World War II
Dates: Opened in 1942
Web: www.muzeum.zagan.pl/en/historia-muzeum (official website)

The area around Zagan in what is now Poland was the site of a number of prisoner of war camps that were used by the Germans during World War II. POWs from countries across Europe were kept here, mostly during the early years of the war. Arguably the most famous of these from an American standpoint was Stalag Luft III, site of the Great Escape, made famous by the Hollywood film of the same name. A museum commemorating all POWs in World War II is now maintained in Zagan.

The first German POW camps were established in Zagan just after Poland fell in 1939. In 1942 Stalag III was opened. Run by the German Air Force, it accommodated Allied airmen, mostly British and Americans, who had been captured after being shot down. A relatively nicer facility than those for common soldiers, detainees here still endeavored on more than one occasion to escape.

The most famous of these attempts was almost a year in the making and involved a tunnel longer than a football field. The escape, which took place on March 24-25, 1944, was initially successful, though many of the escapees were quickly recaptured, and most of these were subsequently executed. The POW camps around Zagan were liberated by the Soviet army in January of 1945. The POW museum was established in Zagan in 1971.

The Museum of the Prisoner of War Camps honors all of those who were held prisoner by the Germans, but focuses mainly on the history of the camps that had been established in Zagan. The museum is actually located in a former POW building, with exhibits and artifacts from the camps as well as some props from the movie *The Great*

Escape. Other monuments in Zagan include a memorial to the fifty POWs who were killed after the escape as well as markers showing where the famous tunnel once ran.

---

229. BIG THREE MONUMENT

Livadiya Park, Yalta, Crimea, Russia

Site Type: Monument – Yalta Conference
Conflict: World War II
Dates: Dedicated in 2015
Web: None Available

The Big Three Monument commemorates the meeting of the three major leaders of the Allied powers in the Crimea in 1945. The Yalta Conference, held in February of that year, saw Joseph Stalin of the Soviet Union, Winston Churchill of the United Kingdom and Franklin Roosevelt of the United States come together to discuss wartime strategy and postwar plans even as Nazi Germany was in its death throes.

After America's entry into World War II, the United States and Great Britain fought the war as full-blown allies, with the Soviet Union as a major partner. The leaders met several times during and immediately after the war, with the most famous conferences taking place at Tehran in 1943 and Yalta in 1945. The Tehran Conference focused primarily on the defeat of Germany, while the Yalta Conference focused on the postwar world order. From a military and political standpoint, these conferences constituted one of the greatest meetings of wartime leaders in history.

The Big Three Monument was erected by Russia in 2015 to commemorate the 70th anniversary of the event. Even though the east and west would spend the next four decades languishing in a Cold War, the alliance was nevertheless a great if brief diplomatic triumph. The massive bronze statue of the three seated leaders can be found in Livadiya Park, not far from the palace where the conference took place.

FOREIGN – SOUTHERN EUROPE & NORTH AFRICA

230. ANZIO BEACHHEAD MUSEUM

Via di Villa Adele 2, 00042 Anzio RM, Italy

Site Type: Museum
Conflict: World War II
Dates: Opened in 1994
Web: www.sbarcodianzio.it (official website)

The Anzio Beachhead Museum is a mid-sized museum which documents one of the most important if lesser known campaigns of the war in Europe: the Battle of Anzio. From late 1942 through 1943 the Western Allies had systematically driven the Axis out of North Africa and Sicily before finally getting bogged down in Southern Italy. The landing at Anzio helped to break the stalemate, and ultimately led to the capture of Rome a few months later.

The Allies began their conquest of Italy in July of 1943. After initial successes in the southern end of the country, the advance became a slog in the face of stiffening Axis resistance. In order to get things moving, the Allies orchestrated an amphibious assault at Anzio behind the German-Italian lines. The initial landings were relatively unopposed, allowing the Allies to establish a strong foothold.

The subsequent campaign, however, quickly ground to a halt. The Germans threw substantial forces against the beachhead, resulting in another stalemate that lasted for four months. It wasn't until May that the Allies gained the upper hand and broke out from the Anzio area. A month later on June 4, just two days before the Normandy invasion, Rome was liberated.

The Anzio Beachhead Museum is located inside an old 17th

century villa in downtown Anzio. While the museum layout is a bit haphazard, it is absolutely jam-packed with fascinating artifacts from the battle. The focus of the collection is on personal items such as uniforms, equipment and photos. All the major combatants are represented, with artifacts from the American, British, Italian and German armies on display.

---

231. SICILY-ROME AMERICAN CEMETERY AND MEMORIAL

Piazzale Kennedy 1, 00048 Neptuno RM, Italy

Site Type: Cemetery
Conflict: World War II
Dates: Opened in 1943
Web: www.abmc.gov/cemeteries-memorials/europe/sicily-rome-american-cemetery (official website)

The Sicily-Rome American Cemetery and Memorial is the largest American military cemetery in Italy and one of the largest in Europe. Many American servicemen who died fighting in Sicily and at Anzio are buried here. It is particularly associated with the latter due to the cemetery's proximity to the Anzio landing site.

The Battle of Anzio witnessed some of the most ferocious fighting during the Italian campaign. An attempt to open the way to Rome for Allied forces working their way up the peninsula, the Anzio landing was ultimately a success, but a pyrrhic one due to the high numbers of casualties on both sides. Unfortunately, after the battle, the Americans moved north to take Rome rather than destroy the surrounded German army, which was given the opportunity to escape and significantly prolong the war in Italy.

The Sicily-Rome American Cemetery was initially established for the burial of those who died fighting at Anzio. Many who were killed in Sicily and elsewhere in Southern Italy were later interred here as well, nearly eight thousand all told. A visitor center with exhibits on the Italian Campaign was added to the memorial in 2014.

232. FLORENCE AMERICAN CEMETERY AND MEMORIAL

Via Cassia, 50023 Tavarnuzze FL, Italy

Site Type: Cemetery
Conflict: World War II
Dates: Established in 1944
Web: www.abmc.gov/cemeteries-memorials/europe/florence-american-cemetery (official website)

The Florence American Cemetery and Memorial is one of the major American cemeteries in Southern Europe. Established in 1944, most of those buried here died while fighting in Northern Italy in the last year of the war.

In the summer of 1944 Rome was liberated by the Allies. This was followed by a swift advance into central Italy and the liberation of Florence. However, since most of the German army had been allowed to retreat when Rome fell, Northern Italy remained solidly under Axis control. Because of this the region saw some of the fiercest fighting in the last months of the war. When Germany finally surrendered, much of Northern Italy was still held by the Axis.

The Florence American Cemetery and Memorial was initially established to accommodate those soldiers who fell in the campaign to overcome the Gothic Line, the last major line of defense of the Germans south of the Alps. Over four thousand American servicemen who died fighting in Northern Italy are buried here.

233. NAVAL MONUMENT AT GIBRALTAR

36 Parliament Lane, GX11 1AA Gibraltar

Site Type: Monument - American Navy in the World Wars
Conflicts: World War I, World War II
Dates: Dedicated in 1937
Web: www.abmc.gov/cemeteries-memorials/europe/naval-monument-gibraltar (official website)

The Naval Monument at Gibraltar was originally built to commemorate the contribution of the American navy to the Allied effort in World War I. It later received an addition commemorating the American landings in North Africa in World War II. It is the largest American military monument in southwestern Europe.

Military activity in the Western Mediterranean during World War I was relatively light, as the bulk of the Central Powers' fleets spent most of the conflict bottled up in home ports. However, the west did see significant enemy submarine activity, and the British/American presence at Gibraltar was critical in checking U-boats from entering the Mediterranean. During World War II Gibraltar became the base where Dwight Eisenhower directed Operation Torch, the American landings in Morocco.

The Naval Monument at Gibraltar, which was built into the existing city wall, was originally completed in 1937. It was constructed using stone quarried from the iconic Rock of Gibraltar. The bronze plaque commemorating Operation Torch was added to the monument in 1998.

234. WESTERN NAVAL TASK FORCE MARKER

Ben M'Sik Cemetery, Boulevard de la Grande Ceinture, Casablanca 20320, Morocco

Site Type: Monument – Battle of Casablanca
Conflict: World War II
Date: Dedicated in 1960 (?)
Web: www.abmc.gov/cemeteries-memorials/africa/western-naval-task-force-marker (official website)

Operation Torch, the invasion of North Africa, was the first American military campaign on land in the European theater of World War II. Fought primarily against Vichy French forces rather than hardened German troops, the main actions of the campaign went relatively smoothly. However, the Vichy forces did put up an unexpected fight at Casablanca.

In November 1942, nearly a year after the United States declared war on Nazi Germany, the first troops were finally ready to go into action. After a trans-Atlantic voyage, landings were made at several points in Morocco and Algeria. Troops of Vichy France, a questionable ally of the Nazis, did resist the landings, forcing the American navy to bombard the landing areas.

A French naval force, despite being outnumbered and outgunned, sailed out to meet the Americans in what was one of the largest surface fleet battles in the Atlantic during the war. Over a dozen French naval vessels, including a cruiser and several destroyers, were sunk, with the American fleet taking minimal damage. Afterwards, the landings at Casablanca proceeded with minimal fuss.

The Western Naval Task Force Marker was erected to commemorate the landing and those who gave their lives liberating Morocco. A small granite monument topped by an American flag, it stands in the Ben M'Sik Cemetery, where many of those Allied Servicemen who died during Operation Torch both on land and sea were buried.

235. NORTH AFRICA AMERICAN CEMETERY AND MEMORIAL

Route de Roosevelt, Carthage, Tunisia

Site Type: Cemetery
Conflict: World War II
Dates: Dedicated in 1960
Web: www.abmc.gov/cemeteries-memorials/africa/north-africa-american-cemetery (official website)

The North Africa American Cemetery and Memorial is the burial site of many American servicemen who died in Tunisian campaign of World War II. Most of these were killed at the Battle of Kasserine Pass in 1943, the first major military engagement of the war fought between the Americans and the Germans.

After landing in Morocco and Algeria in November of 1942 and pushing their way eastward, the American army encountered the first real German resistance in Tunisia. There, along with British troops from Egypt, the Americans fought a series of battles that ultimately ended with the Axis surrender of North Africa along with over two hundred thousand troops. This was the first major victory of the United States in the European theater of the war.

The North Africa American Cemetery and Memorial was established close to where the early fighting took place, near the ruins of the ancient city of Carthage. These ruins were famously depicted in the movie *Patton,* where American General George Patton claims he remembered the Roman siege here from a former life. Nearly three thousand American servicemen are buried at this site.

FOREIGN – FAR EAST

236. UNITED NATIONS MEMORIAL CEMETERY & KOREAN WAR MONUMENT

779-1 Daeyon 4(sa)-dong, Nam-gu, 608-812 Busan, South Korea

Site Type: Monument – Battle of the Pusan Perimeter
Conflict: Korean War
Dates: Battle fought on August 4 – September 18, 1950; Monument dedicated in 2013
Web: www.abmc.gov/cemeteries-memorials/pacific/korean-war-monument-busan (official website)

The United Nations Memorial Cemetery in South Korea is one of the world's most unique military cemeteries. It is the only United Nations cemetery anywhere, serving the dead of all nations who fought to keep South Korea a free and democratic country. Of the seventeen nations that endured combat deaths during the Korean War, ten are represented by the graves here, including the United States. At the center of the cemetery is the Korean War Monument, honoring those Americans who fought and died in the war.

The earliest stages of the Korean War were marked by a string of victories for the Communist North which by August had driven the South Koreans to the southernmost tip of the peninsula. There, with the help of the United States and other allies, they managed to establish a defensive perimeter around the port city of Busan.

For more than a month North Korean forces tried to break through the perimeter in a series of ferocious attacks. However, as time went on, the North Koreans struggled from both dwindling numbers of veteran troops and strained logistics. At the same time supplies and troops from the United States reached Busan, strengthening the defense. After the American landing at Inchon threatened to cut them off, the Communists were forced to withdraw from Busan.

The United Nations Memorial Cemetery was formally established in 1955 by a motion of the General Assembly. However, the first burials of U.N. soldiers began as early as 1950. There are over a dozen monuments here, including a Wall of Remembrance inscribed with the names of over forty thousand dead and missing servicemen. The Korean War Monument specifically commemorates those Americans who fought in the Korean War, especially honoring those who helped to defend the Busan Perimeter at the outset of the conflict.

---

237. LEGATION QUARTER

Dongjiaomin Alley, Dongcheng, 100006 Beijing, China

Site Type: Battlefield – Siege of the International Legations
Conflict: Boxer Rebellion
Dates: Siege fought from June 20-August 14, 1900
Web: www.beijing-visitor.com/beijing-attractions/foreign-legation (official website)

The Siege of the Legation Quarter was one of the most famous military engagements of the early 20th century. It pitted a small international force from a dozen different countries against the Qing-supported Boxers, who sought to liberate Peking, now Beijing, from foreign oppression. The siege lasted fifty-five days and ended with the defeat of the Boxers in Beijing and the flight of the Qing Dynasty from the city.

At the turn of the century the Legation Quarter was home to a number of international diplomatic missions to China, all enclosed within a large compound. The United States was one of the largest of these embassies and contributed a detachment of Marines to the 400-man defense force. The Legation Quarter was a natural target for the Boxers, who, with the reluctant support of the Empress Dowager, sought to kill or drive out all of the city's foreigners in the summer of 1900.

Once the siege began soldiers from the United Kingdom, France, Germany, Austria-Hungary, Italy, Russia, Japan and the United States banded together to defend the compound. Some of the most brutal fighting of the siege took place at the Tartar Wall from June 30-July 3, with the defense of the wall shared by German soldiers and United States Marines. On August 14 a British relief force, followed a few hours later by an American relief force, arrived in Beijing and broke the siege. They occupied the city as the imperial court fled.

The Legation Quarter as it existed in 1900 is now gone, destroyed over the course of several wars and the infamous Cultural Revolution. However, there are still a few old surviving buildings lining Dongjiaomin Alley near Tiananmen Square. Among these is St. Michael's Church, where a separate part of the siege took place, and the old Citibank building, now home to the Beijing Police Museum.

---

238. JIANGSHAN RAIDER MEMORIAL HALL

Bao'an, China

Site Type: Monument – Doolittle Raid
Conflict: World War II
Dates: Dedicated in 2015
Web: None Available

The Jiangshan Raider Memorial Hall is probably among the least well known monuments to Americans in World War II. Sponsored by the Chinese government, it commemorates the Doolittle Raid, a bombing run on Tokyo that was primarily designed to boost American morale following the disaster at Pearl Harbor. The mission was one-way, with the planes and crew escaping to mainland China after the raid.

In the aftermath of Pearl Harbor, the stunned American populace was left with a mix of dread and rage towards the Japanese. Within a few weeks of the attack, the United States military at the behest of Roosevelt began to plan a revenge raid on the Japanese homeland. With few options available in early 1942, it was decided that a small force of bombers would make a surprise bombing run against Tokyo.

The mission was extremely dangerous, as the only way to get bombers in range of Tokyo was by aircraft carrier, and the bombers could not then land again on the carriers. The only choice was to fly to China after the mission was completed and abandon the planes. The mission was indeed a success, and the Japanese were left dumbfounded when American bombs rained down on their capital on April 18, 1942. The bombers continued on to China, where the surviving crew members spent much of the rest of the war.

The Jiangshan Raider Memorial Hall was opened in 2015. It commemorates Jimmy Doolittle, the commander of mission, and the other airmen who served on the raid and who made it to China afterwards. A bronze sculpture out front depicts Doolittle and some of the airmen being assisted by local Chinese citizens.

---

239. HUE WAR MUSEUM

Hai Muoi Ba Thang Tam, Phu Hau, Thanh Pho Hue,
Thura Thien Hue, Vietnam

Site Type: Museum
Conflict: Vietnam War
Dates: Battle fought on January 30-March 3, 1968
Web: www.vietnam.travel/places-to-go/central-vietnam/hue (official tourism website of Hue)

The Battle of Hue was the centerpiece engagement of the Tet Offensive and one of the most important battles of the Vietnam War. Part of an all-out attack across South Vietnam, the effort to dislodge the American forces from their critical base at Hue nearly succeeded, but in the end the city was defended at a terrible cost. Although a victory for the United States and South Vietnam, the ability of North Vietnam to carry out a major offensive coupled with massive casualties helped to turn the tide of American public opinion against the war.

By the beginning of 1968 the war between North and South Vietnam had already been dragging on for well over a decade, with American involvement steadily escalating. After years of stalemate the United States was growing eager to bring the war to a successful conclusion, while the North Vietnamese and their Viet Cong allies were just as eager to prove to the Americans that the war was just getting started. In January of 1968 they launched what is now know as the Tet Offensive, a massive campaign hitting dozens of major targets across the country at once.

The fighting began on the Vietnamese New Year despite a two-day truce which had been agreed to by both sides. After smashing targets across the country, the North Vietnamese turned their attention to the city of Hue on January 31, a highly strategic city and military supply center near the border of the two countries. Over ten thousand North Vietnamese and Vietcong soldiers swarmed in and took most of Hue before the American defense stiffened. Over the next few weeks South Vietnamese reinforcements began to arrive, and Hue was retaken block by block in bitter street fighting. By the beginning of March it was all over.

The Hue War Museum commemorates the battle from the Vietnamese point of view. There are exhibits on the history of the battle, as well as many military artifacts captured from the Americans during the fighting. Other sites of interest throughout the city related to the battle include the bridges over the Perfume River and around the Citadel, the heavily fortified imperial palace on the north side of the river, where some of the most brutal fighting took place.

---

240. KHE SANH COMBAT BASE

Tan Hop, Huong Hoa, Vietnam

Site Type: Battlefield – Battle of Khe Sanh
Conflict: Vietnam War
Date: Battle fought on January 21-July 9, 1968
Web: None Available

The Battle of Khe Sanh was one of the largest military engagements of the Vietnam War. Starting a little over a week before the Tet Offensive, the two campaigns ran concurrent to each other for over two months.

They were in all likelihood both part of one campaign from the point of view of the North Vietnamese, but American historians distinguish the two as separate engagements. While the Battle of Khe Sanh did not have a clear victor, the North Vietnamese came out ahead in its aftermath.

The strategic area in and around the village of Khe Sanh was occupied by American forces in 1964. Located close to the narrowest part of Vietnam, it was used as a base to guard against a North Vietnam attack southward. Naturally it was a prime target for the North Vietnamese during their offensives of early 1968.

A campaign to take the base at Khe Sanh began on January 21. The Tet Offensive began in other places a few days later. The defense of Khe Sanh was fierce and lasted for almost six months. Despite the arrival of additional troops in April, American forces abandoned the base in July. Between Khe Sanh and Tet the North Vietnamese had born extremely heavy casualties, but ultimately gained a strategic advantage that helped to propel them to victory a few years later.

The Khe Sanh Combat Base is a partially preserved American Army base located close to the center of Vietnam. Scattered around the site are remains of trenches and fortifications, as well as several captured helicopters and vehicles. A small museum on site displays captured American weapons and equipment.

---

241. HIROSHIMA PEACE MEMORIAL PARK & MUSEUM

1-2 Nakajimama-Cho, Naka-ku, Hiroshima City 730-08121, Japan

Site Type: Monument – Victims of the Atomic Bomb
Conflict: World War II
Dates: Dedicated in 1954
Designations: UNESCO World Heritage Site
Web: www.hpmmuseum.jp (official website)

The Hiroshima Peace Memorial Park is a large space in the middle of Hiroshima City filled with memorials and monuments, primarily commemorating those who died in the world's first wartime use of an atomic bomb. There are several dozen monuments on the site, plus museums and quiet meditative spaces, honoring many of the different groups of people that suffered in the blast, from children to workers to people in Korea who died later from the fallout.

By 1945 the United States and Japan had been at war for over three years. With the fall of Nazi Germany, America was eager to drive the Japanese to surrender and bring an end to the worst conflict in human history. On August 6 the American bomber Enola Gay dropped an atomic weapon on the city of Hiroshima. The damage was catastrophic. The city was almost completely destroyed, and countless thousands were killed.

This attack, coupled with a similar attack on Nagasaki a few days later, effectively ended the war. In 1954 the Japanese government established a park centered around the location where the bomb detonated. The park was dedicated to the countless citizens who lost their lives in the attack. The Hiroshima Peace Memorial Park was designated a UNESCO World Heritage Site in 1966.

The Hiroshima Peace Memorial Park covers a large area in downtown Hiroshima. There are over sixty memorials, monuments, statues and items of related interest to see. The centerpiece is the A-Bomb Dome, a building close to ground zero that somehow partially survived the blast. Other major memorials include the Children's Peace Monument, the Hiroshima National Peace Memorial Hall, the Atomic Bomb Memorial Mound and the Peace Flame. The Hiroshima Peace Memorial Museum has exhibits on atomic weapons, the bombing, and artifacts that survived the attack.

242. CORNERSTONE OF PEACE

Mabuni, Itoman, Okinawa 901-0333, Japan

Site Type: Monument – Battle of Okinawa
Conflict: World War II
Dates: Dedicated in 1995
Web: http://sp.heiwa-irei-okinawa.jp (official website)

The Cornerstone of Peace monument commemorates the enormous death toll of both soldiers and civilians during the Battle of Okinawa, the last significant campaign of World War II in the Pacific. Following the American conquest of Iwo Jima, Okinawa was the next major step before reaching the Japanese home islands, and its conquest put almost all of Japan within range of American bombers. It was also one of the bloodiest battles of the Pacific War, with over a quarter of a million military and civilian casualties.

In April 1945 the United States began the largest amphibious assault of the Pacific War. While often compared to Iwo Jima, the Okinawa campaign was very different. In addition to the considerably larger military and air forces at the disposal of the Japanese, hundreds of thousands of civilians lived on Okinawa, and a significant percentage of these were drafted to fight in the battle.

The campaign was a slog, lasting for nearly three months before the bulk of the Japanese army was defeated and the island subdued. During this time the United States received over seventy thousand casualties, while Japanese ships and aircraft, including the famous Kamikazis, took a terrible toll on the American navy. In turn, the Japanese took as many as a hundred thousand casualties, with almost all killed. Caught in between, over a hundred thousand civilians, almost half the population of the island, lost their lives.

The Cornerstone of Peace was completed in 1995 on the 50th anniversary of the battle. Inscribed on it are nearly a quarter of a million names of known people who died in the battle, including a huge number of civilians. The Cornerstone of Peace is located in the Okinawa Senseki Quasi-National Park.

243. MOUNT SURIBACHI MEMORIAL

Mount Suribachi, Iwo Jima, Japan

Site Type: Battlefield & Monument – Battle of Iwo Jima
Conflict: World War II
Dates: Battle fought on February 19-March 26, 1945
Web: None Available

The Battle of Iwo Jima was one of the few engagements of World War II which took place on Japanese soil. It is perhaps best known for the enduring legacy of the famous film footage of United States Marines raising the American flag over Mount Suribachi. Defending their home territory for the first time, the Japanese fought without quarter, taking nearly one hundred percent casualties with virtually no one surrendering. The island is now covered with monuments commemorating the battle.

By early 1945, with the Japanese Empire crumbling, the establishment of island bases close to Japan became a prerequisite to an American invasion. The first island to be targeted was Iwo Jima. The Japanese realized both the strategic and symbolic importance of Iwo Jima and built it up into one of the most heavily fortified islands of the empire. Because of this, and because the Japanese were defending their home territory for the first time, the fighting on Iwo Jima was some of the fiercest of the Pacific War.

After months of naval bombardments, tens of thousands of marines began landing on Iwo Jima in February 1945. The casualties in the initial assault were very heavy, and it took the marines four days to secure Mount Suribachi. The American flag was famously raised at the peak of the mountain on February 23. It took a full month to secure the rest of the island. Nearly all of the Japanese defenders were killed, although total American casualties were higher, offering a grim glimpse of what was to be expected in an actual invasion of Japan.

The Mount Suribachi Memorial, located on top of the hill of the same name, commemorates the raising of the American flag here in 1945. Moreover, nearly the entire island of Iwo Jima, which is less than eight square miles in area, is a shrine to the battle. Memorials can be found all over, with markers noting many locations where fighting took place. Some of the Japanese bunkers and tunnels have also been preserved.

---

244. MOUNT SAMAT NATIONAL SHRINE

Mount Samar Road, Pilar, Bataan, Philippines

Site Type: Monument – Battle of Bataan
Conflict: World War II
Date: Battle fought on January 7-April 9, 1942
Web: None Available

The Battle of Bataan was one of the largest military defeats in the history of the United States. Like Pearl Harbor, which was attacked a few weeks before fighting in the Philippines began, Bataan and the infamous death march that followed became a rallying cry in the United States. The Bataan Peninsula was also the location of fierce fighting during the Battle of Luzon in 1945 when the Americans retook the island.

In the wake of the attack on Pearl Harbor the Japanese military began a naval and air version of the Blitzkrieg on American and British territories in the western Pacific. Their primary target was the Philippines. The first attacks were against Luzon Island, home to the colonial capital of Manila and the American Asiatic fleet. The American air forces were taken by surprise and largely destroyed on the first day of the war. Within a week the Japanese had driven off the American navy and secured most of the airfields on Luzon.

Towards the end of December the American commander, Douglas MacArthur, ordered a withdrawal of all troops to the Bataan Peninsula, establishing a series of defensive lines in an effort to delay the Japanese as long as possible. The defense was stubborn and Japanese progress extremely slow. Eventually the defenses started to crumble, and on March 12 MacArthur departed the island. Amazingly, American resistance lasted almost another full month before the final surrender. The survivors were forced to walk the eighty mile Bataan Death March which resulted in horrific casualties.

The Mount Samat National Shrine was erected by the government of the Philippines in honor of the American and Philippine soldiers who fought at the Battle of Bataan. The centerpiece is an enormous memorial cross, one of the largest to be found in the Far East. In addition, there are other monuments scattered throughout Bataan, including the Battle of the Pockets site; the Battle of Layac Junction site; the Fall of Bataan monument; and the Mariveles zero kilometer marker where the Bataan Death March began.

---

245. CLARK VETERANS CEMETERY

Clark Freeport, Mabalacat, Pampanga, Philippines

Site Type: Cemetery
Conflicts: Spanish-American War, World War II
Dates: Opened in 1948
Web: www.abmc.gov/cemeteries-memorials/pacific/clark-veterans-cemetery (official website)

The Clark Veterans Cemetery traces its roots to the turn of the century, when the forces of the United States and Spain clashed near here during the Spanish-American War. Americans who died in that conflict, as well as a large number of Philippine scouts who worked with the United States, were buried in the area that is now Clark Cemetery.

After World War II the graves of a number of American military cemeteries in the Philippines were consolidated here, including most of the Spanish-American war dead and some from World War II. Other American soldiers were later interred here, including numerous unknown soldiers who were killed in later Asian wars.

Clark Veterans Cemetery is located on the site of a former American military base. After its establishment in 1948 thousands of American graves were relocated here from other military cemeteries. Nearly nine thousand servicemen who served in the Spanish-American War and World War II are buried here. In 1991 the cemetery was badly damaged by the eruption of nearby Mount Pinatubo, but has since been restored and reopened to the public.

---

246. CABANATUAN AMERICAN MEMORIAL

Pangatian, Cabanatuan City, Nueva Ecija, Philippines

Site Type: POW Camp
Conflict: World War II
Dates: Established in 1942
Web: www.abmc.gov/cemeteries-memorials/pacific/cabanatuan-american-memorial (official website)

In the wake of the Japanese conquest of the Philippines and the infamous Death March of Bataan which followed, thousands of United States servicemen became POWs. Many of these were interred in, or at least passed through, the prisoner of war camps north of Manila, the most famous of which was at Cabanatuan.

During the war an estimated twenty thousand American and other Allied military personnel were incarcerated at Cabanatuan. Some of these were here for nearly three years. In January of 1945 a joint American-Philipino force liberated the camp in what has famously become known as the Great Raid. More than five hundred prisoners were rescued from the possibility of massacre as the Japanese slowly lost ground on the island.

The Cabanatuan American Monument commemorating the POWs who were kept here was completed in 1982. It stands on part of the former grounds of the camp. Designed as a fenced-in compound reminiscent of the living conditions here during the war, the flags of both the United States and the Philippines fly over a long wall upon which is inscribed the names of over two thousand former prisoners.

247. MANILA AMERICAN CEMETERY AND MEMORIAL

McKinley Road, Fort Bonifacio, Taguig City, Philippines

Site Type: Cemetery
Conflict: World War II
Dates: Opened in 1948
Web: www.abmc.gov/cemeteries-memorials/pacific/manila-american-cemetery (official website)

The Manila American Cemetery is one of the largest American cemeteries outside of the United States, and home to the largest number of graves, at least on foreign soil, of American servicemen from World War II. Many Americans who died in the various Philippines campaigns, as well as other Pacific campaigns, are buried or commemorated here.

During World War II the Philippine islands suffered terribly under two onslaughts: the first by the Japanese in 1941 and 1942, the other by the Americans in 1944 and 1945. Between these two campaigns more than forty thousand servicemen were killed in the Philippines, accounting for more than ten percent of all Americans killed in World War II. Nearly half of these are buried in Manila.

The Manila American Cemetery and Memorial was dedicated in 1948, with many graves consolidated here after the war. Probably the most famous memorial here is that of the Sullivans, five brothers who all served on the USS Juneau and who were all killed on the same day. Their deaths led to a new military policy that split up siblings who were serving in the same units.

248. MACARTHUR LANDING MEMORIAL NATIONAL PARK

Palo, Leyte, Philippines

Site Type: Monument – Douglas MacArthur
Conflict: World War II
Dates: Dedicated in 1977
Designations: Philippines National Park
Web: None Available

On March 12, 1942, with the collapse of American forces in the Philippines imminent, General Douglas MacArthur was ordered to leave the island. According to legend, as he did so, he uttered his now famous phrase "I Shall Return". Interestingly, according to some sources, he may have actually stated this more than a week later after arriving in Australia.

About two and a half years later he did return. On October 20, 1944 MacArthur, who was placed in charge of the Philippines counteroffensive, returned to the Philippines at Leyte. By the end of the war American forces under his command had largely driven the Japanese from the major islands. This campaign cemented MacArthur as one of the great American commanders of the 20th century.

The MacArthur Landing Memorial National Park was created by the government of the Philippines in 1977. It is located within sight of where MacArthur came ashore in 1944. Primarily a protected wildlife habitat, the site includes a monument with seven statues at the water's edge depicting MacArthur and his staff coming ashore. At full tide the monument is partially underwater.

---

249. MARKER AT PAPUA NEW GUINEA

Spring Garden Road, Port Moresby, Papua New Guinea

Site Type: Monument – New Guinea Campaign
Conflict: World War II
Dates: Dedicated in 1992
Web: www.abmc.gov/cemeteries-memorials/pacific/marker-papua-new-guinea (official website)

The Marker at Papua New Guinea commemorates the New Guinea Campaign, the longest nearly continuous campaign of World War II. Begun in January 1942 with the Japanese invasion of the island of Papua, it transitioned over time from an Allied defensive campaign to an offensive one. Fighting continued virtually unabated all the way until the end of the war in 1945.

Within weeks of their crushing victory at Pearl Harbor, Japanese forces fanned out across the Pacific annexing most of the old European colonies. However, the southernmost of the major islands, Papua, was strategically difficult to take quickly due to its size, terrain and its proximity to Australia. Using Australia as a base, Allied forces prevented the Japanese from securing their conquest, essentially making Papua the high water mark of the Japanese Empire.

In 1992 the government of Papua New Guinea honored the United States with a plaque at the American Embassy commemorating the efforts and sacrifice of those Allies who fought to defend Papua from Japan. It was dedicated on the 50th anniversary of Douglas MacArthur's arrival at Port Moresby in 1942.

250. GUADALCANAL AMERICAN MEMORIAL

Skyline Road, Honiara, Solomon Islands

Site Type: Monument – Battle of Guadalcanal
Conflict: World War II
Dates: Battle fought on August 7, 1942-February 9, 1943
Web: www.abmc.gov/cemeteries-memorials/pacific/guadalcanal-memorial (official website)

The Battle of Guadalcanal was the first major American counteroffensive against the Japanese during World War II. This sea, land and air campaign was one of the longest and most comprehensive of the Pacific war, and led for the first time to Japanese loss of territory. It also led to the loss of many irreplaceable Japanese warships that later contributed to the American recovery of island territories.

Within a few months of the attack on Pearl Harbor, Japanese forces occupied most of the major island groups across the Pacific, including the Solomons. The retaking of the Solomon Islands, which stood along the line of communication between the United States and Australia, was a top priority. The first landings took place in August 1942. Within six months, most of the key islands were captured. Naval and aircraft losses were heavy on both sides, but the Japanese forces were irreplaceable, leaving the Americans in a stronger position.

The Guadalcanal American Memorial was inaugurated on the 50th anniversary of the beginning of the battle. It honors all of those Americans who fought and died to liberate the Solomons in 1942 and 1943. Consisting of a series of stone monuments on a wide plaza, the memorial is located on the first hill to be taken by American forces during the battle.

COMING SOON

NATIONAL MUSEUM OF THE UNITED STATES ARMY

1775 Liberty Drive, Fort Belvoir, Virginia, 22060

Site Type: Museum
Dates: Opening in 2020
Web: https://armyhistory.org (official website)

The National Museum of the United States Army is, at the time of this writing, a planned museum currently under construction at Fort Belvoir in Virginia. Sponsored by the United States Army, the Army Historical Foundation and other organizations, it will be one of the largest military museums in the world when completed. The museum is expected to be opened sometime in mid-2020.

The museum is anticipated to have a collection of between forty and fifty thousand pieces, including weaponry, military artifacts, documents and other items. According to their website, a substantial number of the items on display will be rare or unique, including many which will be seen for the first time here.

NATIONAL MEDAL OF HONOR MUSEUM

40 Patriot's Point Road, Mount Pleasant, South Carolina, 29464

Site Type: Museum
Dates: Opening sometime around 2024
Web: www.patriotspoint.org (official website)

The National Medal of Honor Museum is a planned museum which will be located at Patriot's Point outside of Charleston, South Carolina. Sponsored by the National Medal of Honor Museum Foundation and supported by the Congressional Medal of Honor Society, it will be the only officially recognized Medal of Honor museum in United States. The museum is expected to be opened sometime around 2024.

As of the time of this writing, the National Medal of Honor Museum is planned to have exhibits on the history of the medal, its recipients, and the importance of those who have served and died in the cause of freedom.

INDEX OF SITES BY TYPE

BATTLEFIELDS		
5. Bunker Hill Monument	Charlestown, Massachusetts	Battle of Bunker Hill
9. Minute Man National Historical Park	Concord, Massachusetts	Battle of Lexington and Concord
13. Patriot's Park	Portsmouth, Rhode Island	Battle of Rhode Island
14. Great Swamp Fight Monument	West Kingston, Rhode Island	Battle of the Great Swamp
21. White Plains Battlefield Sites	White Plains, New York	Battle of White Plains
23. Bennington Battlefield State Historic Site	Walloomsac, New York	Battle of Bennington
24. Fort William Henry Museum	Lake George, New York	Siege of Fort William Henry
26. Saratoga National Historical Park	Stillwater, New York	Battle of Saratoga
30. Monmouth Battlefield State Park	Manalapan, New Jersey	Battle of Monmouth
32. Trenton Battle Monument	Trenton, New Jersey	Battle of Trenton
40. Gettysburg National Military Park	Gettysburg, Pennsylvania	Battle of Gettysburg
41. Fort Necessity National Battlefield	Farmington, Pennsylvania	Battle of Fort Necessity
42. Woodville Plantation	Bridgeville, Pennsylvania	Battle of Bower Hill
44. Fort McHenry National Monument	Baltimore, Maryland	Battle of Fort McHenry
48. Monocacy National Battlefield	Frederick, Maryland	Battle of Monocacy
50. Antietam National Battlefield	Sharpsburg, Maryland	Battle of Antietam
62. Anacostia Park	Washington, DC	Bonus Army Riot

67. Manassas National Battlefield Park	Manassas, Virginia	1st and 2nd Battles of Bull Run
68. Fredericksburg and Spotsylvania National Military Park	Fredericksburg, Virginia	Battles of Fredericksburg, Chancellorsville, The Wilderness and Spotsylvania
69. Cedar Creek and Belle Grove National Historical Park	Middletown, Virginia	Battle of Cedar Creek
71. Colonial National Historical Park	Yorktown, Virginia	Siege of Yorktown
72. Richmond National Battlefield Park	Richmond, Virginia	Battles of the Peninsula Campaign and Overland Campaign
73. Petersburg National Battlefield	Prince George, Virginia	Siege of Petersburg
77. Appomattox Court House National Historical Park	Appomattox, Virginia	Battle of Appomattox Courthouse
80. Harper's Ferry National Historical Park	Harper's Ferry, West Virginia	Raid on Harper's Ferry
81. Perryville Battlefield State Historic Site	Perryville, Kentucky	Battle of Perryville
85. Perry's Victory and International Peace Memorial	Put-In Bay, Ohio	Battle of Lake Erie
86. Fallen Timbers Battlefield and Fort Miamis National Historic Site	Maumee, Ohio	Battle of Fallen Timbers
88. River Raisin National Battlefield Park	Monroe, Michigan	Battle of Frenchtown
92. Tippecanoe Battlefield Park	Battle Ground, Indiana	Battle of Tippecanoe
98. Moores Creek National Battlefield	Currie, North Carolina	Battle of Moores Creek Bridge
100. Guilford Courthouse National Military Park	Greensboro, North Carolina	Battle of Guilford Courthouse

101. Fort Sumter National Monument	Charleston, South Carolina	Battle of Fort Sumter
105. Kings Mountain National Military Park	Blacksburg, South Carolina	Battle of Kings Mountain
106. Cowpens National Battlefield	Gaffney, South Carolina	Battle of Cowpens
112. Kennesaw Mountain National Battlefield Park	Kennesaw, Georgia	Battle of Kennesaw Mountain
113. Chickamauga and Chattanooga National Military Park	Fort Oglethorpe, Georgia	Battles of Chickamauga and Chattanooga
121. Horseshoe Bend National Military Park	Daviston, Alabama	Battle of Horseshoe Bend
123. Brice's Cross Roads National Battlefield Site	Guntown, Missis-sippi	Battle of Brice's Cross Roads
124. Tupelo National Battlefield	Tupelo, Mississippi	Battle of Tupelo
125. Vicksburg National Military Park	Vicksburg, Missis-sippi	Siege of Vicksburg
128. Stones River National Battlefield	Murfreesboro, Tennessee	Battle of Stones River
129. Franklin Battlefield	Franklin, Tennessee	Battle of Franklin
130. Fort Donelson National Battlefield	Dover, Tennessee	Battle of Fort Donelson
131. Shiloh National Battlefield	Shiloh, Tennessee	Battle of Shiloh
133. Jean Lafitte National Historical Park	Chalmette, Loui-siana	Battle of New Orleans
135. Pea Ridge National Military Park	Garfield, Arkansas	Battle of Pea Ridge
139. Wilson's Creek National Battlefield	Republic, Missouri	Battle of Wilson's Creek
146. Black Jack Battlefield and Nature Park	Wellsville, Kansas	Battle of Black Jack
152. Washita Battlefield National Historic Site	Cheyenne, Okla-homa	Battle of Washita River
154. San Jacinto Battle-ground State Historic Site	La Porte, Texas	Battle of San Jacinto

159. Palo Alto Battlefield National Historical Park	Brownsville, Texas	Battles of Palo Alto and Resaca de la Palma
160. The Alamo Mission	San Antonio, Texas	Siege of the Alamo
165. Sand Creek Massacre National Historic Site	Eads, Colorado	Battle of Sand Creek
174. Little Bighorn Battle-field National Monument	Crow Agency, Montana	Battle of Little Bighorn
175. Big Hole National Battlefield	Wisdom, Montana	Battle of Big Hole
180. Aleutian World War II National Historic Area	Unalaska, Alaska	Battle of Attu
197. USS Arizona Memorial	Honolulu, Hawaii	Battle of Pearl Harbor
199. War in the Pacific National Historical Park	Piti, Guam	Battle of Guam
202. San Juan Hill Battlefield	Santiago de Cuba, Cuba	Battle of San Juan Hill
203. Chapultepec Park	Mexico City, Mexico	Battle of Chapulte-pec
206. Thames Battlefield	Chatham, Canada	Battle of the Thames
210. Normandy American Cemetery and Memorial	Colleville-sur-Mer, France	Battle of Normandy
211. Utah Beach American Memorial and Museum	Ste-Marie-du-Mont, France	Battle of Normandy
216. Belleau Wood American Monument	Belleau, France	Battle of Belleau Wood
217. Chateau-Thierry American Monument	Chateau-Thierry, France	Second Battle of the Marne
218. Bellicourt American Monument	Bellicourt, France	Battle of St. Quentin Canal
219. Meusse-Argonne American Memorial & Montfaucon American Monument	Montfaucon-d'Ar-gonne, France	Battle of Me-usse-Argonne
220. Luxembourg American Cemetery and Memorial	Luxembourg City, Luxembourg	Battle of the Bulge
221. Bastogne War Museum & Mardasson Memorial	Bastogne, Belgium	Siege of Bastogne

224. Hurtgen Forest Battlefield	Hurtgenwald, Germany	Battle of Hurtgen Forest
225. Peace Museum Bridge at Remagen	Remagen, Germany	Battle of the Remagen Bridge
230. Anzio Beachhead Museum	Anzio, Italy	Battle of Anzio
235. North Africa American Cemetery and Memorial	Carthage, Tunisia	Battle of Kasserine Pass
236. United Nations Memorial Cemetery	Busan, South Korea	Battle of the Busan Perimeter
237. Legation Quarter	Beijing, China	Siege of the International Legations
239. Hue War Museum	Thura Thien Hue, Vietnam	Tet Offensive
240. Khe Sanh Combat Base	Huong Hoa, Vietnam	Battle of Khe Sanh
243. Mount Suribachi Memorial	Iwo Jima, Japan	Battle of Iwo Jima
244. Mount Samat National Shrine	Pilar, Philippines	Battle of Bataan
250. Guadalcanal American Memorial	Honiara, Solomon Islands	Battle of Guadalcanal

FORTS		
1. Fort Edgecomb State Historic Site	Edgecomb, Maine	Wooden Blockhouse
2. Fort Knox Historic Site	Prospect, Maine	Masonry Fort
3. Fort at Number 4 Open Air Museum	Charlestown, New Hampshire	Wooden Stockade Fort
16. Fort Trumbull State Park	New London, Connecticut	Masonry Star Fort
17. Castle Clinton National Monument	New York, New York	Masonry Fort
24. Fort William Henry Museum	Lake George, New York	Wooden Star Fort
25. Fort Ticonderoga National Historic Landmark	Ticonderoga, New York	Masonry Star Fort
27. Fort Stanwix National Monument	Rome, New York	Wooden Star Fort

41. Fort Necessity National Battlefield	Farmington, Pennsylvania	Wooden Stockade Fort
43. Fort Delaware State Park	Delaware City, Delaware	Masonry Fort
44. Fort McHenry National Monument	Baltimore, Maryland	Masonry Star Fort
70. Jamestown Settlement Living Museum	Jamestown, Virginia	Wooden Stockade Fort
94. Fort de Chartres State Historic Site	Prairie du Rocher, Illinois	Masonry Fort
99. Fort Raleigh National Historic Site	Manteo, North Carolina	Earthwork Fort
101. Fort Sumter National Monument	Charleston, South Carolina	Masonry Fort
107. Fort Pulaski National Monument	Savannah, Georgia	Masonry Fort
114. Castillo de San Marcos National Monument	St. Augustine, Florida	Masonry Star Fort
118. San Juan National Historic Site	San Juan, Puerto Rico	Walled City
136. Fort Snelling	St. Paul, Minnesota	Masonry Fort
137. Fort Atkinson State Preserve	Fort Atkinson, Iowa	Masonry and Wood Fort
141. Fort Abercrombie State Historic Park	Abercrombie, North Dakota	Wooden Stockade Fort
158. Presidio la Bahia	Goliad, Texas	Masonry Fort (fortified mission)
160. The Alamo Mission	San Antonio, Texas	Masonry Fort (fortified mission)
166. Bent's Old Fort National Historic Site	Le Junta, Colorado	Adobe Fort
169. Cove Fort Historic Site	Beaver, Utah	Masonry Fort
176. Fort Hall	Pocatello, Idaho	Masonry Fort
177. Fort Clatsop	Astoria, Oregon	Wooden Stockade Fort
178. Fort Yamhill Blockhouse	Dayton, Oregon	Wooden Blockhouse
179. Fort Vancouver National Historic Site	Vancouver, Washington	Wooden Stockade Fort

181. Alcatraz Island	San Francisco, California	Masonry Fort
182. Fort Point National Historic Site	San Francsico, California	Masonry Fort
203. Chapultepec Park	Mexico City, Mexico	Masonry Castle
205. Fort Malden National Historic Site	Amherstburg, Canada	Wooden Star Fort

MILITARY BASES, INSTALLATIONS, ACADEMIES & COLLEGES		
15. United States Coast Guard Academy	New London, Connecticut	Military Academy
22. United States Military Academy	West Point, New York	Military Academy
46. United States Naval Academy	Annapolis, Maryland	Military Academy
63. The Pentagon	Washington, DC	Department of Defense HQ
79. Virginia Military Institute	Lexington, Virginia	Military College
82. Fort Knox	Fort Knox, Kentucky	Military Base
102. The Citadel	Charleston, South Carolina	Military College
109. Fort Benning	Columbus, Georgia	Military Base
134. Fort Smith National Historic Site	Fort Smith, Arkansas	Military Base
144. Fort Robinson State Park	Crawford, Nebraska	Military Base
145. Fort Leavenworth	Fort Leavenworth, Kansas	Military Base
147. Fort Scott National Historic Site	Fort Scott, Kansas	Military Base
149. Fort Riley Museums	Fort Riley, Kansas	Military Base
150. Fort Larned National Historic Site	Larned, Kansas	Military Base
151. Fort Sill	Fort Sill, Oklahoma	Military Base
163. Fort Davis National Historic Site	Fort Davis, Texas	Military Base

167. United States Air Force Academy	Colorado Springs, Colorado	Military Academy
168. Cheyenne Mountain Complex	Colorado Springs, Colorado	NORAD Command Center
170. Fort Bowie National Historic Site	Bowie, Arizona	Military Base
171. Davis-Monthan Air Force Base Aircraft Boneyard	Tucson, Arizona	Military Base
172. Fort Churchill State Historic Park	Silver Springs, Nevada	Military Base
173. Fort Laramie National Historic Site	Fort Laramie, Wyoming	Military Base

NAVAL VESSELS		
6. USS Constitution	Charlestown, Massachusetts	Wooden Frigate
10. USS Salem	Quincy, Massachu-setts	Heavy Cruiser
11. USS Massachusetts	Falls River, Massa-chusetts	Battleship
19. USS Intrepid Museum	New York, New York	Aircraft Carrier
31. USS New Jersey	Camden, New Jersey	Battleship
36. USS Olympia	Philadelphia, Pennsylvania	Cruiser
45. USS Constellation	Baltimore, Maryland	Wooden Sloop
75. USS Wisconsin	Norfolk, Virginia	Battleship
95. USS Cobia & Wisconsin Maritime Museum	Manitowoc, Wisconsin	Submarine
96. USS North Carolina	Wilmington, North Carolina	Battleship
103. CSS Hunley & Warren Lasch Conservation Center	North Charleston, South Carolina	Submarine
104. USS Yorktown & Patriot's Point	Mount Pleasant, South Carolina	Aircraft Carrier
119. USS Alabama & Battleship Memorial Park	Mobile, Alabama	Battleship
155. Battleship Texas State Historic Site	La Porte, Texas	Battleship

156. USS Lexington	Corpus Christi, Texas	Aircraft Carrier
183. USS Pampanito & San Francisco Maritime National Historical Park	San Francisco, California	Submarine
186. USS Hornet Museum Ship	Alameda, California	Aircraft Carrier
187. SS Red Oak	Richmond, California	Victory Ship
191. USS Iowa Museum Ship	Los Angeles, California	Battleship
194. USS Midway Museum	San Diego, California	Aircraft Carrier
196. USS Missouri Memorial	Honolulu, Hawaii	Battleship
197. USS Arizona Memorial	Honolulu, Hawaii	Battleship
198. USS Bowfin Submarine Museum	Honolulu, Hawaii	Submarine

MUSEUMS		
10. United States Naval Shipbuilding Museum	Quincy, Massachusetts	Navy
11. Battleship Cove Maritime Museum	Falls River, Massachusetts	Navy
12. Springfield Armory National Historic Site	Springfield, Massachusetts	Weaponry
19. USS Intrepid Sea, Air and Space Museum	New York, New York	Navy, Aviation
28. Buffalo and Erie County Naval and Military Park	Buffalo, New York	Military Vehicles
35. Museum of the American Revolution	Philadelphia, Pennsylvania	American Revolution
36. Independence Seaport Museum	Philadelphia, Pennsylvania	Navy
38. American Military Edged Weaponry Museum	Intercourse, Pennsylvania	Weaponry
39. National Civil War Museum	Harrisburg, Pennsylvania	American Civil War

47. Patuxent River Naval Air Museum	Lexington Park, Maryland	Navy, Aviation
49. National Museum of Civil War Medicine	Frederick, Maryland	Military Medicine, American Civil War
57. National Museum of the United States Navy	Washington, DC	Navy
58. Smithsonian National Air and Space Museum	Washington, DC	Aviation
59. Smithsonian National Museum of American History	Washington, DC	American Military History
60. National Museum of American Jewish Military History	Washington, DC	American Military History
66. National Museum of the Marine Corps	Triangle, Virginia	Navy, Marines
70. Jamestown Settlement Living Museum	Williamsburg, Virginia	Colonial Jamestown
74. United States Army Women's Museum	Fort Lee, Virginia	Army, Women in the Military
76. MacArthur Memorial Museum	Norfolk, Virginia	Douglas MacArthur
78. National Civil War Chaplains Museum	Lynchburg, Virginia	Military Chaplains, American Civil War
82. General George Patton Museum	Fort Knox, Kentucky	George Patton
83. National Museum of the United States Air Force	Dayton, Ohio	Aviation
84. Charles Young Buffalo Soldiers National Monument	Wilberforce, Ohio	Charles Young
89. Yankee Air Museum	Belleville, Michigan	Aviation
90. Indiana World War Memorial Military Museum	Indianapolis, Indiana	World Wars I & II
93. National Veteran's Art Museum	Chicago, Illinois	Military Art
95. Wisconsin Maritime Museum	Manitowoc, Wisconsin	Navy
97. Hannah Block Historic USO Center	Wilmington, North Carolina	USO

103. Warren Lasch Conservation Center	North Charleston, South Carolina	Navy
108. National Museum of the Mighty Eighth Air Force	Pooler, Georgia	Aviation, World War II
109. National Infantry Museum and Soldier Center	Columbus, Georgia	Army
110. National Civil War Naval Museum	Columbus, Georgia	Navy, American Civil War
115. Bay of Pigs Museum	Miami, Florida	Invasion of Cuba
117. National Naval Aviation Museum	Pensacola, Florida	Navy, Aviation
120. Tuskegee Airmen National Historic Site	Tuskegee, Alabama	Aviation, World War II
122. African American Military History Museum	Hattiesburg, Mississippi	African Americans in the Military
127. Sergeant Alvin C. York State Historic Park	Pall Mall, Tennessee	Alvin York
132. National World War II Museum	New Orleans, Louisiana	World War II
140. National World War I Museum and Memorial	Kansas City, Missouri	World War I
148. Dwight D. Eisenhower Presidential Library and Museum	Abilene, Kansas	Dwight Eisenhower
149. Fort Riley Museums	Fort Riley, Kansas	Army
151. United States Army Artillery Museum	Fort Sill, Oklahoma	Army, Weaponry
153. Buffalo Soldiers National Museum	Houston, Texas	Buffalo Soldiers
157. Commemorative Air Force Airpower Museum	Midland, Texas	Aviation
162. National Museum of the Pacific War	Fredericksburg, Texas	World War II
164. White Sands Missile Range Museum	White Sands Missile Range, New Mexico	Weaponry
190. United States Navy Seabee Museum	Port Hueneme, California	Navy
193. San Diego Air and Space Museum	San Diego, California	Aviation

207. American Air Museum in Britain	Duxford, United Kingdom	Aviation
211. Utah Beach American Memorial and Museum	Ste-Marie-du-Mont, France	World War II
212. Memorial Museum of the Battle of Normandy	Bayeux, France	World War II
221. Bastogne War Museum	Bastogne, Belgium	World War II
225. Peace Museum Bridge at Remagen	Remagen, Germany	World War II
226. Allied Museum	Berlin, Germany	Cold War
227. Checkpoint Charlie Museum	Berlin, Germany	Cold War
230. Anzio Beachhead Museum	Anzio, Italy	World War II
239. Hue War Museum	Thura Thien Hue, Vietnam	Vietnam War
241. Hiroshima Peace Memorial Park and Museum	Hiroshima City, Japan	World War II
A. National Museum of the United States Army	Fort Belvoir, Virginia	Army
B. National Medal of Honor Museum	Mount Pleasant, South Carolina	Medal of Honor Recipients

MONUMENTS, MEMORIALS & CEMETERIES		
4. Bennington Battle Monument	Bennington, Vermont	Battle of Bennington
5. Bunker Hill Monument	Charlestown, Massachusetts	Battle of Bunker Hill
7. Massachusetts 54th Regiment Memorial	Boston, Massachu-setts	Massachusetts 54th Regiment
8. Mount Auburn Cemetery	Cambridge, Massachusetts	Julia Ward Howe
17. Battery Park Memorials	New York, New York	Seamen killed in World War II & Korean War Veteans
18. George M. Cohan Monument	New York, New York	George M. Cohan
20. General Grant National Memorial	New York, New York	Ulysses S. Grant

32. Trenton Battle Monument	Trenton, New Jersey	Battle of Trenton
34. Independence National Historical Park	Philadelphia, Pennsylvania	Independence Hall, Liberty Bell, Tomb of the Unknown Revolutionary War Soldier
51. District of Columbia War Memorial	Washington, DC	World War I
52. Navy-Merchant Marine Memorial	Washington, DC	Seaman Killed in World War I
53. National World War II Memorial	Washington, DC	World War II
54. Korean War Veteran's Memorial	Washington, DC	Korean War
55. Vietnam Veteran's Memorial	Washington, DC	Vietnam War
55. Women's Memorial	Washington, DC	Women who served in Vietnam
56. United States Navy Memorial	Washington, DC	United States Navy
61. Congressional Cemetery	Washington, DC	Matthew Brady and John Phillip Sousa
64. Arlington National Cemetery	Arlington, Virginia	Largest Military Cemetery in the United States
65. United States Air Force Memorial	Arlington, Virginia	United States Air Force
84. Charles Young Buffalo Soldiers National Monument	Wilberforce, Ohio	Buffalo Soldiers
85. Perry's Victory and International Peace Memorial	Put-In-Bay, Ohio	Battle of Fallen Timbers
87. Polar Bear Expedition Memorial	Troy, Michigan	Polar Bear Expedition
90. Indiana World War Memorial Plaza	Indianapolis, Indiana	World Wars
91. USS Indianapolis National Memorial	Indianapolis, Indiana	USS Indianapolis

116. Ocala/Marion County Veterans Memorial Park	Ocala, Florida	Navajo Code Talkers
140. National World War I Museum and Memorial	Kansas City, Missouri	World War I
143. Crazy Horse Memorial	Crazy Horse, South Dakota	Crazy Horse
154. San Jacinto Battle-ground State Historic Site	La Porte, Texas	Texas Revolution
161. Military Working Dog Teams National Monument	San Antonio, Texas	Working Dogs
184. West Coast Memorial to the Missing	San Francisco, California	Seamen Killed in World War II
185. Abraham Lincoln Brigade Monument	San Francisco, California	Lincoln Brigade
187. Rosie the Riveter Home Front National Historical Park	Richmond, California	World War II Home Front
188. Port Chicago Naval Magazine National Memorial	Concord, California	Port Chicago Disaster
192. Northwood Gratitude and Honor Memorial	Irvine, California	Wars in Afghanistan and Iraq
195. Bob Hope Memorial	San Diego, California	Bob Hope
197. USS Arizona Memorial	Honolulu, Hawaii	USS Arizona
200. American Memorial Park	Garapan, Northern Mariana Islands	Marianas Campaign
201. Monument to the Victims of the USS Maine	Havana, Cuba	USS Maine
202. Santiago Surrender Tree	Santiago de Cuba, Cuba	Liberation of Cuba
204. Mexico City National Cemetery	Mexico City, Mexico	Mexican-American War
208. Lusitania Memorial Garden	Kinsale, Ireland	Sinking of the Lusitania
209. United States Naval Monument at Brest	Brest, France	American Navy
210. Normandy American Cemetery and Memorial	Colleville-sur-Mer, France	Landings at Omaha Beach

211. Utah Beach American Memorial	Ste-Marie-du-Mont, France	Landings at Utah beach
213. Tours American Monument	Tours, France	Services of Supply
214. Chaumont American Expeditionary Force Headquarters Marker	Chaumont, France	World War I Headquarters
215. Lafayette Escadrille Memorial Cemetery	Marnes-la-Coquette, France	Lafayette Escadrille
216. Belleau Wood American Monument	Belleau, France	Battle of Belleau Wood
217. Chateau-Thierry American Monument	Chateau-Thierry, France	Second Battle of the Marne
218. Bellicourt American Monument	Bellicourt, France	Battle of San Quentin Canal
219. Meusse-Argonne American Memorial & Montfaucon American Monument	Montfaucon-d'Argonne, France	Battle of Meusse-Argonne
220. Luxembourg American Cemetery and Memorial	Luxembourg City, Luxembourg	Battle of the Bulge
221. Mardasson Memorial	Bastogne, Belgium	Siege of Bastogne
222. Ardennes American Cemetery and Memorial	Neupre, Belgium	Battle of the Bulge
223. Netherlands American Cemetery and Memorial	Margraten, Netherlands	Drive to the Rhine
229. Big Three Monument	Yalta, Russia	Big Three Conferences
231. Sicily-Rome American Cemetery and Memorial	Neptuno, Italy	Southern Italian Campaigns
232. Florence American Cemetery and Memorial	Tavarnuzze, Italy	Northern Italian Campaigns
233. Naval Monument at Gibraltar	Gibraltar	American Navy in the World Wars
234. Western Naval Task Force Marker	Casablanca, Morocco	Battle of Casablanca
235. North Africa American Cemetery and Memorial	Carthage, Tunisia	Battle of Kasserine Pass

236. United Nations Memorial Cemetery & Korean War Monument	Busan, South Korea	Battle of Busan Perimeter
238. Jiangshan Raider Memorial Hall	Bao'an, China	Doolittle Raid
241. Hiroshima Peace Memorial Park and Museum	Hiroshima City, Japan	Victims of the Atomic Bomb
242. Cornerstone of Peace	Itoman, Japan	Battle of Okinawa
243. Mount Suribachi Memorial	Iwo Jima, Japan	Battle of Iwo Jima
244. Mount Samat National Shrine	Pilar, Philippines	Battle of Bataan
245. Clark Veterans Cemetery	Mabalacat, Philippines	Philippine Campaigns in the Spanish-American War and World War II
247. Manila American Cemetery	Taguig City, Philippines	Philippine Campaigns in World War II
248. MacArthur Landing Memorial National Park	Palo, Philippines	Douglas MacArthur
249. Marker at Papua New Guinea	Port Moresby, Papau New Guinea	New Guinea Campaign
250. Guadalcanal American Memorial	Honiara, Solomon Islands	Battle of Guadalcanal

OTHER		
111. Andersonville National Historic Site	Andersonville, Georgia	POW Camp
228. Museum of the Prisoner of War Camps	Zagan, Poland	POW Camp
246. Cabanatuan American Memorial	Cabanatuan City, Philippines	POW Camp
189. Tule Lake War Relocation Center	Tulelake, California	Internment Camp
29. Morristown National Historical Park	Morristown, New Jersey	Winter Encampment

37. Valley Forge National Historical Park	King of Prussia, Pennsylvania	Winter Encampment
33. Washington Crossing Parks	Titusville, New Jersey & Washington Crossing, Pennsyl-vania	Northern Campaign
34. Independence National Historic Park	Philadelphia, Pennsylvania	Assorted Sites
126. Manhattan Project National Historical Park	Oak Ridge, Tennessee	Nuclear Research Facility
142. Minute Man Missile National Historic Site	Philip, South Dakota	Ballistic Missile Silo

INDEX OF SITES BY CONFLICT

KING PHILIP'S WAR, KING GEORGE'S WAR & FRENCH AND INDIAN WAR		
3. Fort at Number 4 Open Air Museum	Charlestown, New Hampshire	King George's War, French and Indian War
14. Great Swamp Fight Monument	West Kingston, Rhode Island	King Philip's War
24. Fort William Henry Museum	Lake George, New York	French and Indian War
25. Fort Ticonderoga National Historic Landmark	Ticonderoga, New York	French and Indian War
27. Fort Stanwix National Monument	Rome, New York	French and Indian War
41. Fort Necessity National Battlefield	Farmington, Pennsylvania	French and Indian War

AMERICAN REVOLUTION		
3. Fort at Number 4 open Air Museum	Charlestown, New Hampshire	Northern Campaign
4. Bennington Battle Monument	Old Bennington, Vermont	Northern Campaign
5. Bunker Hill Monument	Charlestown, Massachusetts	Northern Campaign
9. Minute Man National Historical Park	Concord, Massachusetts	Northern Campaign
13. Patriot's Park	Portsmouth, Rhode Island	Northern Campaign
16. Fort Trumbull State Park	New London, Connecticut	Northern Campaign
21. White Plains Battlefield Sites	White Plains, New York	Northern Campaign
23. Bennington Battlefield State Historic Site	Walloomsac, New York	Northern Campaign

25. Fort Ticonderoga National Historic Landmark	Ticonderoga, New York	Northern Campaign
26. Saratoga National Historical Park	Stillwater, New York	Northern Campaign
27. Fort Stanwix National Monument	Rome, New York	Northern Campaign
29. Morristown National Historical Park	Morristown, New Jersey	Northern Campaign
30. Monmouth Battlefield State Park	Manalapan, New Jersey	Northern Campaign
32. Trenton Battle Monument	Trenton, New Jersey	Northern Campaign
33. Washington Crossing Parks	Titusville, New Jersey & Washington Crossing, Pennsylvania	Northern Campaign
34. Independence National Historical Park	Philadelphia, Pennsylvania	Northern Campaign
35. Museum of the American Revolution	Philadelphia, Pennsylvania	Northern Campaign
37. Valley Forge National Historical Park	King of Prussia, Pennsylvania	Northern Campaign
71. Colonial National Historical Park	Yorktown, Virginia	Southern Campaign
98. Moores Creek National Battlefield	Currie, North Carolina	Southern Campaign
100. Guilford Courthouse National Military Park	Greensboro, North Carolina	Southern Campaign
105. Kings Mountain National Military Park	Blacksburg, South Carolina	Southern Campaign
106. Cowpens National Battlefield	Gaffney, South Carolina	Southern Campaign

BARBARY WARS, CITIZEN UPRISINGS & EARLY AMERICAN INDIAN WARS		
6. USS Constitution	Charlestown, Massachusetts	Barbary Wars
46. Tripoli Monument	Annapolis, Maryland	Barbary Wars
12. Springfield Armory National Historic Site	Springfield, Massachusetts	Shay's Rebellion

42. Woodville Plantation	Bridgeville, Pennsylvania	Whiskey Rebellion
86. Fallen Timbers Battlefield and Fort Miamis National Historic Site	Maumee, Ohio	Northwest Indian War
90. Tippecanoe Battlefield Park	Battle Ground, Indiana	Tecumseh's War
121. Horseshoe Bend National Military Park	Daviston, Alabama	Creek War
134. Fort Smith National Historic Site	Fort Smith, Arkansas	Trail of Tears

WAR OF 1812		
1. Fort Edgecomb State Historic Site	Edgecomb, Maine	War of 1812
6. USS Consitution	Charlestown, Massachusetts	War of 1812
44. Fort McHenry National Monument	Baltimore, Maryland	War of 1812
85. Perry's Victory Monument and International Peace Memorial	Put-In-Bay, Ohio	War of 1812
88. River Raisin National Battlefield Park	Monroe, Michigan	War of 1812
121. Horseshoe Bend National Military Park	Daviston, Alabama	War of 1812
133. Jean Lafitte National Historical Park	Chalmette, Louisiana	War of 1812
205. Fort Malden National Historic Site	Amherstburg, Canada	War of 1812
206. Thames Battlefield	Chatham, Canada	War of 1812

TEXAS REVOLUTION & MEXICAN-AMERICAN WAR		
154. San Jacinto Battleground State Historic Site	La Porte, Texas	Texas Revolution
158. Presidio La Bahia	Goliad, Texas	Texas Revolution

160. The Alamo Mission	San Antonio, Texas	Texas Revolution
159. Palo Alto Battlefield National Historical Park	Brownsville, Texas	Mexican-American War
166. Bent's Old Fort National Historic Site	Le Junta, Colorado	Mexican-American War
203. Chapultepec Park	Mexico City, Mexico	Mexican-American War
204. Mexico City National Cemetery	Mexico City, Mexico	Mexican-American War

AMERICAN CIVIL WAR & RELATED CONFLICTS		
146. Black Jack Battlefield and Nature Park	Wellsville, Kansas	Bleeding Kansas
147. Fort Scott National Historic Site	Fort Scott, Kansas	Bleeding Kansas
80. Harper's Ferry National Historical Park	Harper's Ferry, West Virginia	Harper's Ferry Raid
7. Massachusetts 54th Regiment Memorial	Boston, Massachusetts	Civil War
8. Mount Auburn Cemetery	Cambridge, Massachusetts	Civil War
20. General Grant National Memorial	New York, New York	Civil War
39. National Civil War Museum	Harrisburg, Pennsylvania	Civil War
40. Gettysburg National Military Park	Gettysurg, Pennsylvania	Civil War
43. Fort Delaware State Park	Delaware City, Delaware	Civil War
45. USS Constellation	Baltimore, Maryland	Civil War
48. Monocacy National Battlefield	Frederick, Maryland	Civil War
49. National Museum of Civil War Medicine	Frederick, Maryland	Civil War
50. Antietam National Battlefield	Sharpsburg, Maryland	Civil War
67. Manassas National Battlefield Park	Manassas, Virginia	Civil War

68. Fredericksburg and Spotsylvania National Military Park	Fredericksburg, Virginia	Civil War
69. Cedar Creek and Belle Grove National Historical Park	Middletown, Virginia	Civil War
72. Richmond National Battlefield Park	Richmond, Virginia	Civil War
73. Petersburg National Battlefield	Prince George, Virginia	Civil War
77. Appomattox Court House National Historical Park	Appomattox, Virginia	Civil War
78. National Civil War Chaplain's Museum	Lynchburg, Virginia	Civil War
81. Perryville Battlefield State Historic Site	Perryville, Kentucky	Civil War
101. Fort Sumter National Monument	Charleston, South Carolina	Civil War
103. CSS Hunley & Warren Lasch Conservation Center	North Charleston, South Carolina	Civil War
107. Fort Pulaski National Monument	Savannah, Georgia	Civil War
110. National Civil War Naval Museum	Columbus, Georgia	Civil War
111. Andersonville National Historic Site	Andersonville, Georgia	Civil War
112. Kennesaw Mountain National Battlefield Park	Kennessaw, Georgia	Civil War
113. Chickamauga and Chattanooga National Military Park	Fort Oglethorpe, Georgia	Civil War
123. Brice's Cross Roads National Battlefield Site	Guntown, Mississippi	Civil War
124. Tupelo National Battlefield	Tupelo, Mississippi	Civil War
125. Vicksburg National Military Park	Vicksburg, Mississippi	Civil War
128. Stones River National Battlefield	Murfreesboro, Tennessee	Civil War

129. Franklin Battlefield	Franklin, Tennessee	Civil War
130. Fort Donelson National Battlefield	Dover, Tennessee	Civil War
131. Shiloh National Military Park	Shiloh, Tennessee	Civil War
134. Fort Smith National Historic Site	Fort Smith, Arkansas	Civil War
135. Pea Ridge National Military Park	Garfield, Arkansas	Civil War
139. Wilson's Creek National Battlefield	Republic, Missouri	Civil War

LATER AMERICAN INDIAN WARS		
141. Fort Abercrombie State Historic Site	Abercrombie, North Dakota	Sioux Wars
143. Crazy Horse Memorial	Crazy Horse, South Dakota	Sioux Wars
144. Fort Robinson State Park	Crawford, Nebraska	Sioux Wars
150. Fort Larned National Historic Site	Larned, Kansas	Sioux Wars
173. Fort Laramie National Historic Site	Fort Laramie, Wyoming	Sioux Wars
174. Little Bighorn Battlefield National Monument	Crow Agency, Montana	Sioux Wars
152. Washita Battlefield National Historic Site	Cheyennee, Oklahoma	Colorado War
165. Sand Creek Massacre National Historic Site	Eads, Colorado	Colorado War
163. Fort Davis National Historic Site	Fort Davis, Texas	Apache Wars
170. Fort Bowie National Historic Site	Bowie, Arizona	Apache Wars
172. Fort Churchill State Historic Park	Silver Springs, Nevada	Pyramid Lake War
175. Big Hole National Battlefield	Wisdom, Montana	Nez Perce War

SPANISH-AMERICAN WAR & BOXER REBELLION		
36. USS Olympia	Philadelphia, Pennsylvania	Spanish-American War
201. Monument to the Victims of the USS Maine	Havana, Cuba	Spanish-American War
202. San Juan Hill Battlefield & Santiago Surrender Tree	Santiago de Cuba, Cuba	Spanish-American War
245. Clark Veterans Cemetery	Mabalacat, Philippines	Spanish-American War
237. Legation Quarter	Beijing, China	Boxer Rebellion

WORLD WAR I & INTERWAR PERIOD		
18. George M. Cohan Monument	New York, New York	World War I
36. USS Olympia	Philadelphia, Pennsylvania	World War I
51. District of Columbia War Memorial	Washington, DC	World War I
52. Navy-Merchant Marine Memorial	Washington, DC	World War I
127. Sergeant Alvin C. York State Historic Park	Pall Mall, Tennessee	World War I
140. National World War I Museum and Memorial	Kansas City, Missouri	World War I
155. Battleship Texas State Historic Site	La Porte, Texas	World War I
208. Lusitania Memorial Garden	Kinsale, Ireland	World War I
209. United States Naval Monument at Brest	Brest, France	World War I
213. Tours American Monument	Tours, France	World War I
214. Chaumont American Expeditionary Force Headquarters Marker	Chaumont, France	World War I
215. Lafayette Escadrille Memorial Cemetery	Marnes-la-Coquette, France	World War I
216. Belleau Wood American Monument	Belleau, France	World War I

217. Chateau-Thierry American Monument	Chateau-Thierry, France	World War I
218. Bellicourt American Monument	Bellicourt, France	World War I
219. Meusse-Argonne American Memorial & Montfaucon American Monument	Montfaucon-d'Argonne, France	World War I
233. Naval Monument at Gibraltar	Gibraltar	World War I
87. Polar Bear Expedition Memorial	Troy, Michigan	Russian Civil War
62. Anacostia Park	Washington, DC	Bonus Army Riot
185. Abraham Lincoln Brigade Monument	San Francisco, California	Spanish Civil War

WORLD WAR II		
17. Battery Park Memorials	New York, New York	European Theater
120. Tuskegee Airmen National Historic Site	Tuskegee, Alabama	European Theater
155. Battleship Texas State Historic Sitre	La Porte, Texas	European Theater (primarily)
210. Normandy American Cemetery and Memorial	Colleville-sur-Mer, France	European Theater
211. Utah Beach American Memorial and Museum	Ste-Marie-du-Mont, France	European Theater
212. Memorial Museum of the Battle of Normandy	Bayeux, France	European Theater
220. Luxembourg American Cemetery and Memorial	Luxembourg City, Luxembourg	European Theater
221. Bastogne War Museum & Mardasson Memorial	Bastogne, Belgium	European Theater
222. Ardennes American Cemetery and Memorial	Neupre, Belgium	European Theater
223. Netherlands American Cemetery and Memorial	Margraten, Netherlands	European Theater

224. Hurtgen Forest Battlefield	Hurtgenwald, Germany	European Theater
225. Peace Museum Bridge at Remagen	Remagen, Germany	European Theater
228. Museum of the Prisoner of War Camps	Zagan, Poland	European Theater
229. Big Three Monument	Yalta, Russia	European Theater
230. Anzio Beachhead Museum	Anzio, Italy	European Theater
231. Sicily-Rome American Cemetery and Memorial	Neptuno, Italy	European Theater
232. Florence American Cemetery and Memorial	Tavarnuzze, Italy	European Theater
233. Naval Monument at Gibraltar	Gibraltar	European Theater
234. Western Naval Task Force Marker	Casablanca, Morocco	European Theater
235. North Africa American Cemetery and Memorial	Carthage, Tunisia	European Theater
11. USS Massachusetts	Falls River, Massachusetts	Pacific Theater (primarily)
19. USS Intrepid Sea, Air and Space Museum	New York, New York	Pacific Theater
31. USS New Jersey	Camden, New Jersey	Pacific Theater
75. USS Wisconsin	Norfolk, Virginia	Pacific Theater
91. USS Indianapolis National Memorial	Indianapolis, Indiana	Pacific Theater
95. USS Cobia & Wisconsin Maritime Museum	Manitowoc, Wisconsin	Pacific Theater
96. USS North Carolina	Wilmington, North Carolina	Pacific Theater
104. USS Yorktown & Patriot's Point	Mount Pleasant, South Carolina	Pacific Theater
119. USS Alabama & Battleship Memorial Park	Mobile, Alabama	Pacific Theater (primarily)
156. USS Lexington	Corpus Christi, Texas	Pacific Theater
162. National Museum of the Pacific War	Fredericksburg, Texas	Pacific Theater

180. Aleutian World War II National Historic Area	Unalaska, Alaska	Pacific Theater
183. USS Pampanito & San Francisco Maritime National Historical Park	San Francisco, California	Pacific Theater
184. West Coast Memorial to the Missing	San Francisco, California	Pacific Theater
186. USS Hornet Museum Ship	Alameda, California	Pacific Theater
191. USS Iowa Museum Ship	Los Angeles, California	Pacific Theater
196. USS Missouri Memorial	Honolulu, Hawaii	Pacific Theater
197. USS Arizona Memorial	Honolulu, Hawaii	Pacific Theater
198. USS Bowfin Submarine Museum	Honolulu, Hawaii	Pacific Theater
199. War in the Pacific National Historical Park	Piti, Guam	Pacific Theater
200. American Memorial Park	Garapan, Northern Mariana Islands	Pacific Theater
238. Jiangshan Raider Memorial Hall	Bao'an, China	Pacific Theater
241. Hiroshima Peace memorial Park and Museum	Hiroshima City, Japan	Pacific Theater
242. Cornerstone of Peace	Itoman, Japan	Pacific Theater
243. Mount Suribachi Memorial	Iwo Jima, Japan	Pacific Theater
244. Mount Samat National Shrine	Pilar, Philippines	Pacific Theater
245. Clark Veterans Cemetery	Mabalacat, Philippines	Pacific Theater
246. Cabanatuan American Memorial	Cabanatuan City, Philippines	Pacific Theater
247. Manila American Cemetery	Taguig City, Philippines	Pacific Theater
248. MacArthur Landing Memorial National Park	Palo, Philippines	Pacific Theater
249. Marker at Papau New Guinea	Port Moresby, Papau New Guinea	Pacific Theater
250. Guadalcanal American Memorial	Honiara, Solomon Islands	Pacific Theater

53. National World War II Memorial	Washington, DC	General
97. Hannah Block Historic USO Center	Wilmington, North Carolina	General
126. Manhattan Project National Historical Park	Oak Ridge, Tennessee	General
132. National World War II Museum	New Orleans, Louisiana	General
157. Commemorative Air Force Airpower Museum	Midland, Texas	General
187. Rosie the Riveter Home Front National Historical Park	Richmond, California	General
188. Port Chicago Naval Magazine National Memorial	Concord, California	General
189. Tule Lake War Relocation Center	Tulelake, California	General

KOREAN WAR, VIETNAM WAR & COLD WAR EVENTS		
17. Battery Park Memorials	New York, New York	Korean War
54. Korean War Veteran's Memorial	Washington, DC	Korean War
75. USS Wisconsin	Norfolk, Virginia	Korean War
191. USS Iowa Museum Ship	Los Angeles, California	Korean War
196. USS Missouri Memorial	Honolulu, Hawaii	Korean War
198. USS Bowfin Submarine Museum	Honolulu, Hawaii	Korean War
236. United Nations Memorial Cemetery & Korean War Monument	Busan, South Korea	Korean War
31. USS New Jersey	Camden, New Jersey	Korean War, Vietnam War
19. USS Intrepid Sea, Air and Space Museum	New York, New York	Vietnam War
55. Vietnam Veteran's Memorial & Women's Memorial	Washington, DC	Vietnam War
104. USS Yorktown & Patriot's Point	Mount Pleasant, South Carolina	Vietnam War

186. USS Hornet Museum Ship	Alameda, California	Vietnam War
194. USS Midway Museum	San Diego, California	Vietnam War
239. Hue War Museum	Thura Thien Hue, Vietnam	Vietnam War
240. Khe Sanh Combat Base	Huong Hoa, Vietnam	Vietnam War
10. USS Salem	Quincy, Massachusetts	Cold War
115. Bay of Pigs Museum	Miami, Florida	Cold War
142. Minuteman Missile National Historic Site	Philip, South Dakota	Cold War
156. USS Lexington	Corpus Christi, Texas	Cold War
168. Cheyenne Mountain Complex	Colorado Springs, Colorado	Cold War
226. Allied Museum	Berlin, Germany	Cold War
227. Checkpoint Charlie Museum	Berlin, Germany	Cold War

MIDDLE EAST WARS		
191. USS Iowa Museum Ship	Los Angeles, California	Iran-Iraq War
192. Northwood Gratitude and Honor Memorial	Irvine, California	Afghanistan War, Iraq War
194. USS Midway Museum	San Diego, California	Iraq War
196. USS Missouri Memorial	Honolulu, Hawaii	Gulf War